ESSENTIAL OIL CHEMISTRY: UNLOCKING THE SECRETS OF AROMATIC COMPOUNDS

REBECCA PARK TOTILO

AROMAHUT
INSTITUTE

Copyright

Essential Oil Chemistry: Unlocking the Secrets of Aromatic Compounds

This book's purpose is to educate laypersons and health practitioners on the safe use of essential oils and clinical aromatherapy. It is offered with the understanding that Rebecca at the Well Foundation and/or Aroma Hut Institute or Rebecca Park Totilo shall not have any liability or responsibility for any injury caused directly or indirectly by the information contained in this course. Every effort has been taken to provide accurate information, however, none of the information in this course should be construed as medical advice or treatment. Please seek medical advice and recommendations for appropriate healthcare from a qualified medical healthcare provider if needed.

Aroma Hut Institute
https://AromaHut.com

ISBN: 979-8-9872464-6-7

Table of Contents

Table of Contents

04

05

06

Table of Contents

Aromatherapy, as a discipline, rests on a foundation of intricate chemical interactions.

CHAPTER ONE

Introduction

Essential oils represent a complex and intriguing facet of natural product chemistry. Their use spans centuries and cultures, from ancient healing practices to modern holistic health applications. This book is designed to provide a comprehensive exploration of the chemistry underpinning essential oils, aimed at students and practitioners pursuing an advanced understanding of aromatherapy.

Aromatherapy, as a discipline, rests on a foundation of intricate chemical interactions. Each essential oil comprises a diverse array of chemical constituents, each contributing to the oil's unique aromatic profile and therapeutic properties. A profound grasp of these chemical elements is indispensable for any serious aromatherapist.

This book will guide you through the sophisticated world of essential oil chemistry, beginning with the fundamental principles of organic chemistry before progressing to more intricate topics. We will delve into the primary chemical families prevalent in essential oils—such as terpenes, alcohols, esters, and aldehydes—and scrutinize their individual and synergistic effects. Additionally, we will explore advanced extraction techniques and the variables that influence the quality, potency, and efficacy of essential oils. What is an esential oil?

Mastering the chemistry of essential oils enables a more nuanced and effective application in therapeutic practices. Whether you are formulating blends for specific therapeutic outcomes or creating bespoke aromatic compositions,

> *The primary constituents of essential oils are aromatic compounds that give each oil its unique scent and therapeutic properties.*

this chemical knowledge is an essential asset. By deepening your scientific understanding, you will be better prepared to harness the therapeutic power of these extraordinary natural compounds and excel in the field of aromatherapy.

What Is an Essential Oil?

An essential oil is a concentrated, volatile, aromatic liquid extracted from various parts of plants, including leaves, flowers, stems, roots, seeds, or bark. These oils capture the plant's scent and flavor, or "essence," and are often referred to as the plant's life force or immune system due to their potent chemical properties. Essential oils are characterized by their volatility, meaning they readily evaporate at room temperature, allowing them to disperse their fragrance into the air quickly. The primary constituents of essential oils are aromatic compounds that give each oil its unique scent and therapeutic properties.

Extraction Methods

The extraction methods for essential oils vary and include steam distillation, cold pressing, solvent extraction, and CO2 extraction. Steam distillation is the most common method, involving steam passed through plant material to vaporize the volatile compounds, which are then condensed back into liquid form. Cold pressing is typically used for citrus fruits, where the oil is mechanically pressed out of the rinds. Solvent extraction is employed for delicate plant materials that cannot withstand heat, using solvents to dissolve the essential oils, which are then separated from the solvent. CO2 extraction uses supercritical carbon dioxide to extract the oils, preserving a broader range of the plant's aromatic compounds.

The extraction method and production process significantly impact the chemistry of an essential oil. Different extraction techniques can alter the chemical composition, affecting the oil's therapeutic properties and overall quality.

Steam distillation, the most widely used method, generally preserves most components well, but the high temperatures can cause thermal degradation of heat-sensitive compounds. Cold pressing, commonly used for citrus oils, preserves the full spectrum of volatile compounds since it does not involve heat; however, it may also include more non-volatile residues. Solvent extraction, used for delicate flowers like jasmine and tuberose, can leave behind solvent residues and extract both volatile and non-volatile components, including waxes and pigments, which might not be desirable. CO_2 extraction, utilizing supercritical carbon dioxide, can extract a broader range of compounds without high temperatures, resulting in oils with richer and more complex profiles. Additionally, this method leaves no solvent residue, maintaining the essential oil's purity. Hydrodistillation, similar to steam distillation but involving direct contact of water with plant material, often results in a slightly different chemical profile. Maceration, where plant materials are soaked in a carrier oil, infuses the carrier oil with the plant's aromatic compounds, resulting in a product that is not a pure essential oil but a blend that includes components from both the plant and the carrier oil.

Factors such as temperature, pressure, and duration of the extraction process play critical roles in determining the final chemical composition of essential oils. High temperatures can degrade heat-sensitive components, while pressure can influence which compounds are

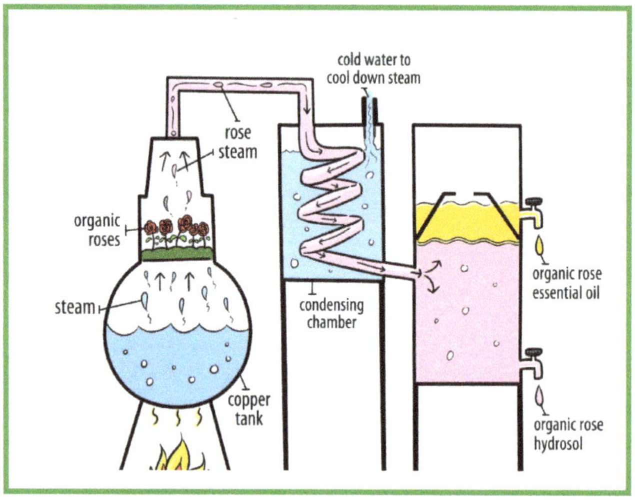

Longer extraction times can lead to the breakdown of certain compounds, altering the oil's chemistry.

extracted. Longer extraction times can lead to the breakdown of certain compounds, altering the oil's chemistry. Solvents can extract different components, thus changing the oil's chemical profile. Moreover, the freshness and quality of the plant material can affect the final chemical composition.

Thermal degradation is a notable concern, particularly for monoterpenes, which can break down under high heat. Oxidation, due to exposure to air and light during extraction, can also alter the oil's chemistry. Solvent extraction methods may leave traces of the solvent, impacting the essential oil's chemical composition and potentially therapeutic efficacy.

Understanding these factors is crucial for ensuring the desired quality and therapeutic properties of the essential oils produced or used in aromatherapy. The extraction method must be carefully considered to maintain the integrity of the essential oil's chemical profile.

Extraction Methods and Their Impact

Does the extraction method and production of an essential oil change the chemistry of an oil? Yes, the extraction method and production process can significantly impact the chemistry of an essential oil. Here are some key points to consider:

STEAM DISTILLATION

Common Method: Most widely used for extracting essential oils.
Impact on Chemistry: The high temperatures can cause thermal degradation of heat-sensitive compounds. However, it generally preserves most of the oil's components well.

COLD PRESSING

Common for Citrus Oils: Used mainly for extracting oils from citrus peels.
Impact on Chemistry: This method preserves the full spectrum of volatile compounds, as it does not involve heat. However, it may include more non-volatile residues.

SOLVENT EXTRACTION

For Delicate Flowers: Used for jasmine and tuberose, which are too delicate for steam distillation.
Impact on Chemistry: Solvents can leave behind residues and may extract both volatile and non-volatile components, including waxes and pigments, which might not be desirable.

CO2 EXTRACTION

Supercritical CO2: Uses supercritical carbon dioxide as a solvent.
Impact on Chemistry: This method can extract a broader range of compounds without using high temperatures, resulting in oils with richer and more complex profiles. It also leaves no solvent residue.

HYDRO DISTILLATION

Similar to Steam Distillation: The plant material is immersed in water and then boiled.
Impact on Chemistry: Like steam distillation, it often results in a slightly different chemical profile due to the water's direct contact with the plant material.

MACERATION

Infusion: Plant materials are soaked in a carrier oil.
Impact on Chemistry: This method infuses the carrier oil with the plant's aromatic compounds but results in a product that is not a pure essential oil. The chemical composition includes components from both the plant and the carrier oil.

Factors Affecting Essential Oil Chemistry

Temperature: High temperatures can cause the degradation of heat-sensitive components.

Pressure: In methods like steam distillation, pressure can influence which compounds are extracted.

Duration: Longer extraction times can result in the breakdown of certain compounds.

Solvents Used: Different solvents can extract different components, altering the oil's chemical profile.

Plant Material Quality: The freshness and quality of the plant material can affect the final chemical composition.

Specific Changes in Chemistry

Thermal Degradation: Some components, such as monoterpenes, can break down under high heat.

Oxidation: Exposure to air and light during extraction can lead to oxidation of certain compounds, altering the oil's chemistry.

Solvent Residue: Solvent extraction methods can leave behind traces of the solvent, which can change the oil's chemical composition.

Understanding these factors is crucial for ensuring the desired quality and therapeutic properties of the essential oils you produce or use.

Steam distillation is the most widely used method for extracting essential oils like lavender, peppermint, and eucalyptus. It utilizes steam to release and separate the oils from the plant material.

An essential oil is a concentrated, volatile extract from plants that captures the essence of their aroma and therapeutic properties.

CHAPTER TWO

Molecular Structures and Functional Groups

Understanding essential oil components' molecular structures and functional groups is fundamental to grasping their chemical behavior and therapeutic properties. These aspects form the foundation of essential oil chemistry, revealing how different molecules interact with biological systems and contribute to the oils' diverse effects.

The makeup of an essential oil begins with a fundamental review of chemistry. Chemistry is the study of chemicals made of numerous elements, separated into two types: organic and inorganic. Because it was believed the element carbon was only produced by living things, the study of carbon was called "organic chemistry." In contrast, other elements such as helium, iodine, and uranium were placed into the "inorganic chemistry" category. And since not all carbon reactions are organic, another way to look at organic chemistry would be to consider the study of molecules containing the carbon-hydrogen (C-H) bond and their reactions.

An essential oil's constituents are, in fact, molecules, which are structured in a unique way that provides them with their aroma and therapeutic benefits. Like all other living matter, essential oils contain chemical compounds of oxygen, carbon, and hydrogen, which create a synergy of different constituents. Therefore, since essential oils are composed of mainly carbon compounds and are a collection of molecules such as carbon, oxygen, and hydrogen, the study of essential oils falls under the "organic chemistry" category.

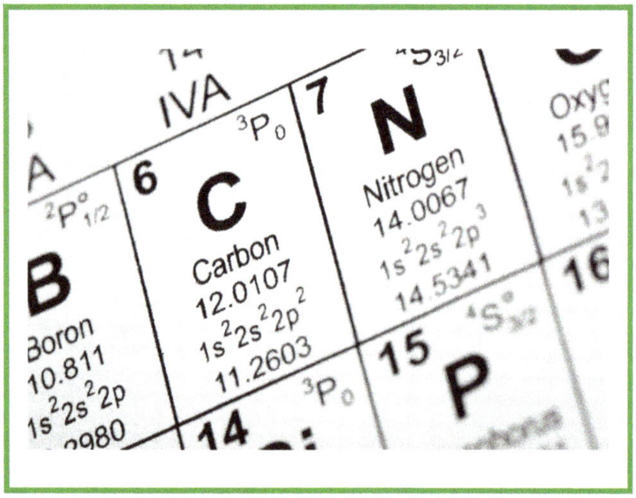

Molecules are comprised of the chemical constituents responsible for the oils' therapeutic properties.

What Is a Molecule?

A molecule is the smallest unit of a chemical compound that retains its chemical properties. It consists of two or more atoms bonded through chemical bonds, which can be covalent (sharing of electrons), ionic (transferring electrons), or metallic. Molecules can range in complexity from simple diatomic molecules like oxygen (O_2) and nitrogen (N_2) to complex organic compounds such as proteins and DNA.

In the context of essential oils and aromatherapy, molecules are crucial as they comprise the chemical constituents responsible for the oils' therapeutic properties. Each molecule's structure and functional groups determine its specific effects and interactions with biological systems.

Every element that makes up a compound has been identified and given a name (with an abbreviation). The main elements discussed in this study include carbon, hydrogen, oxygen, and sometimes nitrogen and sulfur. The abbreviations for these elements are the following:

- Carbon (C)
- Hydrogen (H)
- Oxygen (O)
- Nitrogen (N)
- Sulphur (S)

So, every essential oil, such as lavender, balsam fir, or rose, has unique carbon compound ingredients. When you bend down to smell a rose in your garden, the vapor entering your nostrils is a unique combination of carbon compounds your brain has learned to identify as a rose.

As you can probably guess, there are infinite combinations of how carbon molecules and hydrogen atoms arrange themselves. Depending on where oxygen bonds in the carbon chain, you will get entirely different molecules once you add oxygen into the mix. All of these factors determine the chemical properties of an oil and how they work.

What Is a Hydrocarbon?

A hydrocarbon is an organic compound composed exclusively of hydrogen and carbon atoms. These compounds are the simplest form of organic molecules and serve as the foundational building blocks for more complex substances. Hydrocarbons are classified based on the types of bonds between carbon atoms: alkanes (single bonds), alkenes (at least one double bond), alkynes (at least one triple bond), and aromatic hydrocarbons (ring structures with alternating double bonds). Hydrocarbons such as terpenes in essential oils are crucial in defining their aromatic properties and therapeutic effects. Hydrocarbons thus significantly contribute to the efficacy and versatility of essential oils in aromatherapy.

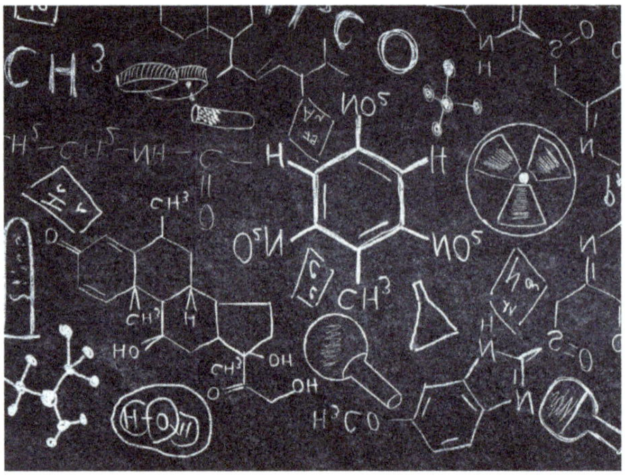

Terpenes are the main constituents of essential oils, responsible for their aromas and therapeutic effects.

What Is a Terpene?

While every essential oil is made from several different molecules and compounds, you will find that they all have a carbon-hydrogen backbone. One of the most familiar components you will work with is terpenes.

A terpene is a molecule (or chemical constituent) made of carbon and hydrogen. Terpenes comprise various isoprene units, including monoterpenes, sesquiterpenes, and diterpenes. Almost one hundred percent of the chemical components that make up all essential oils include terpenes and terpenoid compounds, sesquiterpenes, sesquiterpenoid compounds, and phenylpropane derivatives. Terpineols are formed from acetyl-coenzyme A, which plays a vital role in the body's production of hormones, vitamins, and energy. In addition, terpenes can inhibit the accumulation of toxins and help discharge toxins from the body, including the liver and kidneys.

Terpenes are the primary constituents of essential oils and are responsible for their distinctive aromas and therapeutic properties. Terpenes are classified based on the number of isoprene units they contain. The three major groups of terpenes are monoterpenes, sesquiterpenes, and diterpenes. Each group has unique chemical structures, properties, and applications in aromatherapy and other fields. These groups of terpenes will be discussed in the following sections.

Monoterpenes

A monoterpene is made up of ten carbon atoms. It can take on a variety of shapes, incorporating double bonds to produce a variety of different terpenes. Monoterpenes consist of two isoprene units and have the molecular formula C10H16. Monoterpenes are highly volatile organic compounds with lower boiling points, making them ideal for extraction through steam distillation.

These compounds are prevalent in various essential oils, each contributing unique aromas and therapeutic properties. For example, citrus oils such as lemon and orange are rich in limonene, which provides a fresh, zesty aroma and offers antioxidant and anti-inflammatory benefits. Pine oil, containing high levels of **pinene**, has a distinctive pine scent and is known for its anti-inflammatory, antimicrobial, and bronchodilator effects. Lavender oil, abundant in **linalool**, boasts a floral, sweet aroma and is renowned for its calming, anxiolytic, and analgesic properties. Geranium oil, rich in **geraniol**, has a sweet, rosy scent and demonstrates antimicrobial, antioxidant, and anti-inflammatory effects. Bay oil, containing significant amounts of **myrcene**, features an earthy, herbal fragrance known for its analgesic, anti-inflammatory, and sedative effects. **Camphene**, known for its pungent and earthy aroma and its anti-inflammatory and antioxidant effects, is found in camphor, rosemary, and fir oils. These monoterpenes not only define the characteristic scents of these essential oils but also enhance their therapeutic efficacy, making steam distillation a preferred method for their extraction.

Sesquiterpenes

A sesquiterpene has fifteen carbon atoms and can take on various forms with a molecular structure of C15H24. Like monoterpenes, sesquiterpenes may be acyclic or contain rings, including many unique combinations. Composed of three isoprene units, sesquiterpenes are a significant class of compounds found in many essential oils, known for their therapeutic properties and complex aromas. With their higher molecular weights and boiling points, these compounds are less volatile than monoterpenes but still effectively extracted through steam distillation.

Through steam distillation, the chemistry of rose essential oil captures over 300 unique compounds, including geraniol and citronellol, delivering potent therapeutic benefits.

Patchouli oil (Pogostemon cablin) contains patchoulol, a sesquiterpene that gives the oil its deep, musky fragrance and exhibits antimicrobial and anti-inflammatory properties.

For instance, ginger oil (*Zingiber officinale*) is rich in **zingiberene**, contributing to its warm, spicy aroma and anti-inflammatory, analgesic, and antioxidant properties. Chamomile oil (*Matricaria recutita*) contains significant amounts of **bisabolol**, known for its calming, anti-inflammatory, and antimicrobial effects. Ylang ylang oil (*Cananga odorata*) is abundant in **farnesene**, which imparts a sweet, floral scent and offers anti-inflammatory and sedative benefits. Patchouli oil (*Pogostemon cablin*) contains **patchoulol**, a sesquiterpene that gives the oil its deep, musky fragrance and exhibits antimicrobial and anti-inflammatory properties. Sandalwood oil (*Santalum album*) is rich in **santalol**, which provides a woody, earthy aroma and has been shown to possess anti-inflammatory, antimicrobial, and calming effects. These sesquiterpenes not only define the unique scents of these essential oils but also enhance their therapeutic benefits, making them valuable in aromatherapy and natural medicine.

Ginger oil, rich in zingiberene, offers a warm, spicy aroma along with powerful anti-inflammatory, analgesic, and antioxidant benefits.

Diterpenes

Diterpenes have twenty carbon atoms and four isoprene units. However, this group rarely survives the steam distillation process, so to extract diterpenes, other methods such as solvent extraction, CO2 extraction, or enfleurage are more effective. Several essential oils contain diterpenes. Rosemary (*Rosmarinus officinalis*) essential oil is rich in diterpenes, particularly **rosmarinic acid** and **carnosic acid**. These compounds are known for their antioxidant and anti-inflammatory properties. Sage (*Salvia officinalis*) essential oil contains various diterpenes, including **carnosic acid** and **carnosol**. These compounds have demonstrated antioxidant, antimicrobial, and anti-inflammatory effects. Patchouli (*Pogostemon cablin*) essential oil contains diterpenes such as **patchoulol** and **pogostol**. These compounds contribute to patchouli oil's distinct aroma and potential therapeutic properties, including antimicrobial and anti-inflammatory effects. Sandalwood (*Santalum album*) essential oil contains diterpenes like **santalols**, responsible for their unique fragrance and potential medicinal properties, including anti-inflammatory and antimicrobial effects. Cedarwood (*Cedrus spp.*) essential oil, derived from various species of cedarwood trees, contains diterpenes such as **cedrol**. These compounds are known for their woody aroma and potential insecticidal properties. Fennel (*Foeniculum vulgare*) essential oil contains diterpenes like **fenchone** and **estragole**. Fennel oil is known for its sweet, licorice-like aroma and potential digestive and antimicrobial properties.

Aromatic Compounds

In essential oils, aromatic compounds are organic molecules characterized by their stable ring structures containing alternating double bonds, known as aromatic rings. These compounds are known for their distinct and often pleasant fragrances, contributing significantly to the scent profiles of essential oils. A common example of an aromatic compound in essential oils is **benzene**, which serves as a building block for more complex aromatic structures.

Aromatic compounds in essential oils are important not only for their olfactory properties but also for their therapeutic benefits. These compounds can exhibit various biological activities, such as antimicrobial, anti-inflammatory, and antioxidant effects. For instance, **eugenol**, found in clove oil, is an aromatic compound with well-documented antiseptic and analgesic properties. Another example is **thymol**, present in thyme oil, known for its potent antimicrobial and antifungal activities.

Aromatic compounds in essential oils enhance their effectiveness in aromatherapy, providing both sensory pleasure and health benefits. These compounds' stability and specific chemical interactions make them valuable components in the formulation and application of essential oils for therapeutic purposes.

Aromatic compounds in essential oils are valued not only for their scent but also for their therapeutic benefits, exhibiting biological activities like antimicrobial, anti-inflammatory, and antioxidant effects.

Molecular Structures

The molecular structure of a compound refers to the arrangement of atoms within the molecule and the chemical bonds that hold these atoms together. In essential oils, these structures can range from simple to highly complex, influencing the volatility, solubility, and reactivity of the compounds. The backbone of many essential oil molecules is based on isoprene units, leading to the formation of terpenes and their derivatives. Understanding molecular structures helps us predict how essential oil components behave under various conditions, such as exposure to light, heat, or air, which can affect their stability and efficacy.

Functional Groups

Functional groups are specific groupings of atoms within molecules that have characteristic properties and reactivities. These groups define the chemical behavior of the molecule and play a crucial role in determining the therapeutic actions of essential oils. Typical functional groups in essential oils include alcohols, aldehydes, ketones, esters, ethers, phenols, and acids. Each functional group imparts distinct properties, such as antimicrobial activity, anti-inflammatory effects, or calming aromas.

For example, alcohols (e.g., linalool in lavender) are generally known for their antimicrobial and soothing properties, while aldehydes (e.g., citral in lemongrass) are recognized for their strong, often citrusy scents and calming effects. Phenols (e.g., thymol in thyme) are potent antimicrobials but require careful handling due to their potential skin-irritating properties.

Chemical Families

In chemistry, a "chemical family" refers to a group of chemical compounds that share similar structural features, properties, and chemical behavior due to their similar chemical composition. Chemical families are sometimes called "chemical groups" or "chemical series." These families are often defined by a common functional group or by the elements and bonds that make up their core structure. A few examples of well-known chemical families include terpenes, which include monoterpenes and sesquiterpenes, which are prevalent in essential oils and known for their diverse aromatic profiles and therapeutic effects. Alcohols, such as **linalool** and **menthol**, are characterized by their soothing and anti-inflammatory properties. Esters, including **linalyl acetate** and **geranyl acetate**, are recognized for their pleasant, fruity aromas and calming effects. Aldehydes like **citral** and **cinnamaldehyde** offer strong, distinctive odors and can provide antimicrobial and stimulating benefits. Ketones, such as **carvon** and **menthone**, have therapeutic properties but may require cautious use due to potential irritant effects. Finally, phenols like **thymol** and **eugenol** are noted for their antiseptic and analgesic properties. Each chemical family contributes unique qualities to essential oils, influencing their scent and therapeutic applications.

Chemical families are useful in organic chemistry because they help chemists predict the properties and reactivity of compounds within the same family. They share similar chemical reactions and often have analogous chemical behavior due to their common structural elements. This knowledge is essential for understanding and manipulating organic compounds in various chemical processes.

Common Chemical Families in Essential Oils

Essential oils are complex mixtures of various chemical compounds, and they can be categorized into several chemical families based on the predominant types of compounds they contain. Here is a list of some common chemical families found in essential oils:

Terpenes: Terpenes are essential oils' most prominent and most diverse family of compounds. They include compounds like limonene, pinene, myrcene, and terpinolene. Essential oils rich in terpenes often have a fresh, citrusy, or pine-like aroma. Examples include lemon (limonene), pine (pinene), and lavender (linalool) essential oils.

Monoterpenes: Monoterpenes are a subgroup of terpenes and are characterized by their simple molecular structure. They are often found in oils derived from citrus fruits and coniferous trees. Essential oils rich in monoterpenes include orange (limonene), cypress (pinene), and basil (eucalyptol).

Sesquiterpenes: Sesquiterpenes are larger terpenes with more complex structures. They are commonly found in essential oils derived from roots, woods, and resins. Examples include cedarwood (cedrol), frankincense (α-pinene, β-caryophyllene), and ginger (zingiberene) essential oils.

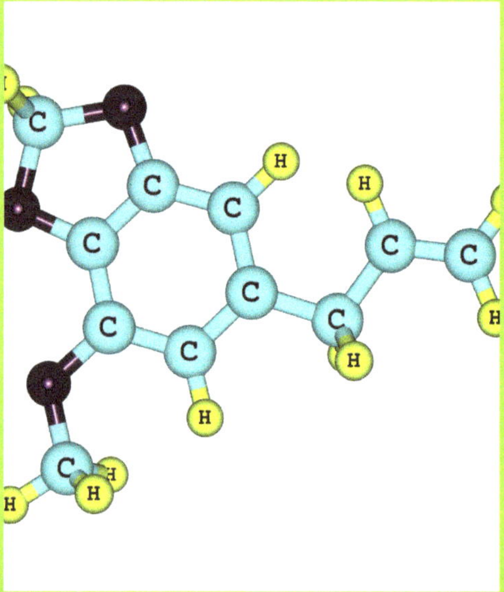

Phenols: Phenols are compounds with a hydroxyl group (-OH) attached to an aromatic ring. They have strong antiseptic and disinfectant properties. Examples of essential oils rich in phenols include oregano (carvacrol), thyme (thymol), and clove (eugenol).

Aldehydes: Aldehydes are compounds characterized by the presence of the carbonyl group (-CHO). They often have a fruity or citrus-like aroma. Citronellal in citronella oil and citral in lemongrass oil are examples of aldehydes in essential oils. Lemongrass (citral), citronella (citronellal), and cassia (cinnamaldehyde) are other examples.

Ketones: Ketones contain a carbonyl group within the molecule. They are found in oils like peppermint and sage and often have a camphoraceous scent. Peppermint (menthone), sage (thujone), and helichrysum (β-diketone) are essential oils that contain ketones.

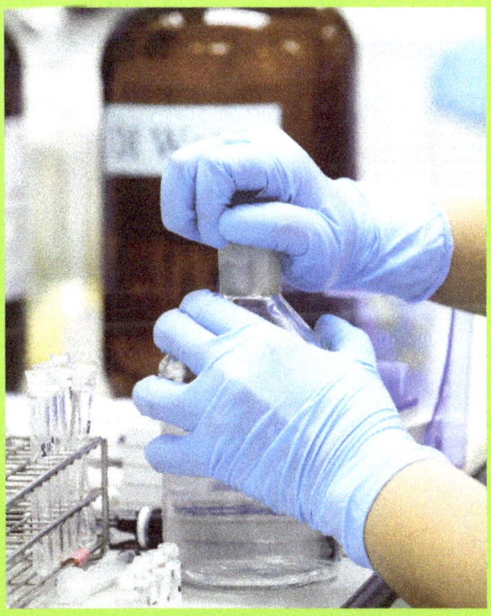

Esters: Esters are compounds formed from the reaction between an alcohol and an organic acid. Esters have a specific structure, with a carbonyl group (C=O) bonded to an oxygen atom and an organic group bonded to the oxygen atom. Ethyl acetate (CH3COOC2H5) is a common ester. They are known for their sweet, fruity, and floral aromas. Examples include linalyl acetate in lavender and bergamot and geranyl acetate in geranium.

Oxides: Oxides contain an oxygen atom within a ring structure. Eucalyptol, found in eucalyptus oil, is a common oxide in essential oils and has a fresh, camphor-like scent. Eucalyptus (1,8-cineole),

tea tree (terpinen-4-ol), and ravintsara (1,8-cineole) are others.

Alcohols: Alcohols are compounds with a hydroxyl group (-OH) attached to a carbon atom. They are prevalent in essential oils like lavender and tea tree and often have a pleasant, slightly floral aroma. Other essential oils include rose (citronellol), tea tree (terpinen-4-ol), and sandalwood (santalol).

Thiols: Thiols are sulfur-containing compounds found in oils like garlic and onion. They have a strong, pungent odor and are less common in aromatherapy due to their intense scent. Garlic (allicin) and onion (propanethiol) essential oils are examples.

Lactones: Lactones are cyclic esters, and they contribute to the creamy, coconut-like scent of oils like ylang ylang (γ-lactones), sweet marjoram (lactones), and verbena (verbascoside).

Coumarins: Coumarins are aromatic compounds with a sweet, hay-like aroma. They are found in oils like sweet woodruff and sweet clover. Sweet woodruff (coumarin), sweet clover (dicoumarol), and Tonka Bean (coumarin) are a few essential oils that contain coumarins.

Furanoids: Contributing to their fragrance and therapeutic properties, but can cause phototoxic reactions. Examples of these in essential oils include angelica root (angelicin) and grapefruit (bergamottin).

Ethers: Known for their distinctive ether linkage (R-O-R), ethers contribute to an oil's

antispasmodic, anesthetic, and anti-inflammatory properties. Examples of essential oils include anise (anethole), fennel (anethole), and basil (methyl chavicol).

Acids: Though they are less common, they contribute to the oils' therapeutic properties, such as antimicrobial, anti-inflammatory, and astringent effects, with examples including acetic acid and benzoic acid. Examples of essential oils include valerian (valeric acid) and rose (citric acid).

Nitrogen: This is characterized by the presence of nitrogen atoms within their molecular structure. These compounds can contribute to the oils' therapeutic properties and distinct aromas. Some examples of essential oils containing nitrogen compounds include chamomile (*Matricaria chamomilla* and *Chamaemelum nobile*), which contains azulene, a nitrogenous compound, particularly in the form of chamazulene. yarrow (*Achillea millefolium*) contains proazulenes that convert to azulenes upon distillation, which are nitrogenous compounds, and mustard (*Brassica spp.*) contains allyl isothiocyanate, a nitrogen-containing compound responsible for the pungent aroma. Others include black pepper (piperine) and marjoram (octen-3-ol).

Chemical Compounds

A chemical compound in an essential oil refers to a specific, well-defined molecule composed of two or more atoms of different elements bonded together, such as terpenes like limonene, phenols, or thymol, which are responsible for the oil's distinct fragrance

"A chemical compound in an essential oil, such as limonene or thymol, is a distinct molecule that gives the oil its unique fragrance and health benefits.

and potential health benefits. These compounds can be isolated and identified through chemical analysis.

These elements are bonded through chemical bonds, and the resulting compound has unique properties different from its constituent elements. Chemical compounds have a specific chemical formula, which indicates the types and proportions of atoms in the compound. For example, water (H_2O) is a chemical compound consisting of two hydrogen (H) atoms and one oxygen (O) atom. Chemical compounds are typically pure substances with distinct chemical and physical properties. Various analytical techniques like spectroscopy and chromatography can characterize them.

In organic chemistry, the terms "compound" and "component" describe different aspects of chemical substances and their structures. A compound is a substance that consists of two or more other elements chemically bonded together in fixed proportions. Chemical bonds, such as covalent bonds, typically hold these elements together. Compounds have distinct and unique chemical properties that differ from their constituent elements' properties. This means that compounds often exhibit entirely new characteristics, such as different melting points, boiling points, reactivity, and chemical behavior.

Steam distillation preserves most essential oil compounds, though high heat can degrade heat-sensitive components.

Common Compounds in Essential Oils

Essential oils contain various compounds, each contributing to their unique properties and therapeutic benefits. Here are a few examples of common compounds found in essential oils.

These compounds contribute to the therapeutic qualities of essential oils and are often responsible for their characteristic aromas.

COMPOUNDS	ESSENTIAL OILS FOUND IN	PROPERTIES
Linalool	Lavender, coriander, bergamot, basil	Calming, sedative, anti-inflammatory, analgesic
Limonene	Citrus oils (lemon, orange, lime, grapefruit)	Antioxidant, anti-inflammatory, mood-enhancing, immune-boosting
Menthol	Peppermint, eucalyptus	Cooling, analgesic, anti-inflammatory, decongestant
Eugenol	Clove, cinnamon, basil	Antiseptic, analgesic, anti-inflammatory, antioxidant
Citral	Lemongrass, lemon myrtle, lemon tea tree	Antimicrobial, anti-inflammatory, sedative, antioxidant
Geraniol	Rose, palmarosa, citronella, gernaium	Antioxidant, anti-inflammatory, antimicrobial, insect-repellent
Thymol	Thyme, oregano	Antimicrobial, antifungal, antioxidant, anti-inflammatory
Camphor	Camphor tree, rosemary	Stimulant, decongestant, anti-inflammatory, analgesic

Common Compounds in Essential Oils

COMPOUNDS	ESSENTIAL OILS FOUND IN	PROPERTIES
1,8-Cineole (Eucalyptol)	Eucalyptus, rosemary, tea tree	Expectorant, decongestant, antimicrobial, anti-inflammatory
Carvacrol	Oregano, thyme	Antimicrobial, antifungal, antioxidant, anti-inflammatory
Menthylamine	Peppermint	Digestive aid, muscle relaxant, headache relief, anti-inflammatory, antimicrobial, cooling effect
Anethole	Fennel	Carminative, digestive aid, antispasmodic, expectorant, antimicrobial
Citric Acid	Citrus oils (lemon, lime, grapefruit)	Antimicrobial, astringent, preservative

Essential oils combine compounds from various chemical families, giving them unique aromas and benefits.

Chemical Components

A component refers to any individual element or group of atoms within a chemical compound. Components can be either atoms or functional groups, which are specific arrangements of atoms with distinct chemical properties. They are the building blocks of compounds, determining the compound's chemical structure and properties. The arrangement and connectivity of components within a compound are crucial for understanding its behavior and reactivity. For example, in the essential oil of lavender, the key components include carbon (C), hydrogen (H), and the hydroxyl functional group (-OH) found in linalool. Each component contributes to the oil's overall structure and properties, influencing its aromatic profile and therapeutic effects.

In summary, a compound is a substance formed by the chemical bonding of two or more elements. At the same time, a component refers to the individual elements or groups of atoms within that compound. Understanding both compounds and their components is fundamental in organic chemistry, as it helps chemists analyze and manipulate the properties and reactivity of various substances.

Essential oils comprise various chemical components, each contributing to their unique properties and therapeutic effects. The chart below gives a few examples of key components commonly found in essential oil.

Common Components in Essential Oils

These components are responsible for essential oils' diverse effects and therapeutic benefits, making them valuable in aromatherapy and natural health practices.

COMPONENTS	EXAMPLES	ESSENTIAL OILS FOUND IN	PROPERTIES
Monoterpenes	Limonene, pinene	Citrus oils (limonene in lemon, orange), pine (pinene)	Antimicrobial, anti-inflammatory, uplifting, immune-boosting
Sesquiterpenes	Bisabolol, caryophyllene	German chamomile (bisabolol), black pepper (caryophyllene)	Anti-inflammatory, calming, analgesic, skin-healing
Monoterpenols	Linalool, geraniol	Lavender (linalool), rose (geraniol)	Antiseptic, calming, balancing, antimicrobial
Sesquiterpenols	Santalol, cedrol	Sandalwood (santalol), cedarwood (cedrol)	Grounding, relaxing, antimicrobial, anti-inflammatory
Phenols	Thymol, eugenol	Thyme (thymol), clove (eugenol)	Strongly antiseptic, stimulating, antioxidant
Aldehydes	Citral, neral	Lemongrass (citral), lemon balm (neral)	Antimicrobial, anti-inflammatory, calming, sedative
Ketones	Camphor, thujone	Rosemary (camphor), sage (thujone)	Respiratory support, mucolytic, wound-healing
Oxides	1,8-Cineole (eucalyptol)	Eucalyptus, rosemary	Expectorant, decongestant, antimicrobial, anti-inflammatory

Common Components in Essential Oils

COMPONENTS	EXAMPLES	ESSENTIAL OILS FOUND IN	PROPERTIES
Esters	Linalyl acetate, geranyl acetate	Lavender (linalyl acetate), sweet marjoram (geranyl acetate)	Calming, anti-inflammatory, antispasmodic, balancing
Lactones	Costunolide	Costus root, elecampane	Anti-inflammatory, antimicrobial, antispasmodic
Acids	Citric acid, myrtanic acid	Citrus oils (citric acid), myrtle (myrtanic acid)	Antimicrobial, astringent, preservative

A chemical constituent refers to the individual chemical compounds present within the oil, each of which contributes to its unique aroma, therapeutic properties, and overall composition.

Chemical Constituents

A chemical constituent is a specific substance or compound part of a larger chemical mixture or composition. In essential oils, a chemical constituent refers to the individual chemical compounds present within the oil, each of which contributes to its unique aroma, therapeutic properties, and overall composition. Essential oils comprise various chemical constituents, including terpenes, alcohols, phenols, esters, and other organic compounds. Each essential oil has a unique composition that gives it its characteristic aroma and potential therapeutic properties. Chemical constituents in essential oils are not typically pure compounds; they are a blend of multiple compounds, each contributing to the oil's overall scent and potential benefits.

In summary, a chemical compound is a pure substance formed by the chemical combination of elements, while a chemical constituent in an essential oil refers to the various organic compounds present in the oil mixture, which collectively give the essential oil its distinct properties and fragrance. Essential oils are complex mixtures of these constituents and not single chemical compounds.

Remember that many essential oils contain a combination of compounds from different chemical families, which gives them their unique aromas and therapeutic properties. In addition, Aromatic compounds, particularly phenylpropanoids like eugenol (found in clove oil) and cinnamaldehyde (found in cinnamon oil), fall under these broader categories. They are termed "aromatic" because of their pleasant smells and the presence of an aromatic ring in their structure.

It's important to understand the chemical composition of essential oils when using them for aromatherapy or other applications to ensure safe and effective use. By understanding these additional families and subcategories, you can gain a more comprehensive view of the chemical diversity present in essential oils. Specific essential oil chemistry references and textbooks can provide more in-depth information if you need further detailed classification.

Chemical Families

FAMILY	COMPONENTS
Terpenes	Hydrocarbons consisting of isoprene units. • Monoterpenes (e.g., Limonene, Pinene) • Sesquiterpenes (e.g., Bisabolol, Farnesene)
Terpenoids	Modified terpenes with added functional groups. • Monoterpenoids (e.g., Linalool, Geraniol) • Sesquiterpenoids (e.g., Chamazulene, Farnesol)
Phenylpropanoids	Compounds derived from phenylpropane (e.g., Eugenol, Anethole)
Acids	Organic compounds with a carboxyl group (e.g., Citric acid, Myrtanic acid)
Ketones	Organic compounds with a carbonyl group (e.g., Carvone, Menthone)
Aldehydes	Organic compounds with a formyl group (e.g., Citral, Benzaldehyde)
Alcohols	Organic compounds with a hydroxyl group (e.g., Linalool, Geraniol)
Esters	Organic compounds formed from alcohol and acid (e.g., Linalyl acetate, Geranyl acetate)
Oxides	Compounds containing an oxygen atom bonded to other elements (e.g., 1,8-Cineole, Ascaridole)
Ethers	Compounds with an oxygen atom connected to two alkyl or aryl groups (e.g., Methyl chavicol, Myristicin)

Chemical Types

TYPES	DESCRIPTION
Hydrocarbons	Compounds composed only of hydrogen and carbon. • Monoterpenes (Limonene, Pinene), Sesquiterpenes (Bisabolol, Farnesene)
Oxygenated Compounds	Compounds that contain oxygen atoms in addition to hydrogen and carbon. • Alcohols (Linalool, Geraniol), Aldehydes (Citral, Benzaldehyde), Ketones (Carvone, Menthone), Esters (Linalyl acetate, Geranyl acetate)
Carboxylic Acids	Organic acids containing a carboxyl group. • Citric acid, Myrtanic acid
Other	Compounds that do not fit neatly into the above categories. • Oxides (1,8-Cineole, Ascaridole), Ethers (Methyl chavicol, Myristicin)

Chemical Components

FAMILY	CONSTITUENTS
Terpenes	**Monoterpenes:** Limonene, Pinene **Sesquiterpenes:** Bisabolol, Farnesene
Terpenoids	**Monoterpenoids:** Linalool, Geraniol **Sesquiterpenoids:** Chamazulene, Farnesol
Phenylpropanoids	Eugenol, Anethole
Acids	Citric acid, Myrtanic acid
Ketones	Carvone, Menthone
Aldehydes	Citral, Benzaldehyde
Alcohols	Linalool, Geraniol
Esters	Linalyl acetate, Geranyl acetate
Oxides	1,8-Cineole, Ascaridole
Ethers	Methyl chavicol, Myristicin

Chemical Compounds

COMPOUNDS
Refers to the specific chemical substances found in essential oils, which can be categorized into both types and components. For example, limonene is a compound that is a monoterpene (Type: hydrocarbon, Family: terpene) and also a specific component.

Components in Different Families

Each component has a specific chemical structure that determines its classification within a particular chemical family. This classification helps in understanding their properties, therapeutic effects, and interactions in aromatherapy and essential oil chemistry.

FAMILIES	COMPONENTS
Terpenes	**Monoterpenes**: Limonene, Pinene, Myrcene **Sesquiterpenes**: Farnesene, Bisabolene
Terpenoids	**Monoterpenoids**: Linalool, Geraniol, Citral **Sesquiterpenoids**: Farnesol, Nerolidol
Phenylpropanoids	Eugenol, Anethole, Safrole
Ketones	Carvone, Menthone, Thujone
Aldehydes	Citral, Cinnamaldehyde, Benzaldehyde
Alcohols	Linalool, Geraniol, Terpineol
Esters	Linalyl acetate, Geranyl acetate, Methyl salicylate
Oxides	1,8-Cineole (Eucalyptol), Ascaridole
Ethers	Anethole, Methyl chavicol (Estragole)
Acids	Citric acid, Myrtanic acid, Valerenic acid

”

Understanding chemical families and their components is crucial for mastering the diverse properties, therapeutic benefits, and applications of essential oils.

CHAPTER THREE

Understanding the Chemistry

Understanding an essential oil's components can help you assess its suitability for treating a specific condition. For example, it is helpful to use oils that contain azulene or bisabolol to treat inflammation, as these have recognized anti-inflammatory qualities. Therefore, determining the chemistry of essential oils is vital when creating an aromatherapy blend. This knowledge can also help you devise a treatment plan—for example, whether you require robust and short-term or milder treatment over a more extended period. Certain chemicals are suited for working in different ways.

However, it is important to realize that aromatherapy is not purely based on chemical analysis—essential oils are intricate and are not a simple product of their base ingredients. Geranium is known to have certain qualities that its known chemical ingredients cannot explain—the overall synergy of constituents is what ultimately determines an essential oil's unique character.

When studying essential oil components, it is imperative to consider the oil as a whole. Some chemicals may only be responsible for a tiny percentage of the oil's composition yet play a large part in its unique character. An excellent example is B-damascenone, which makes up only 0.14% of rose essential oil but is responsible for its distinctive rose fragrance. Kurt Schnaubelt, Ph.D., author of *Advanced Aromatherapy*, states that the classification system based on chemical components of oils is the most effective way to determine which oils to use in an essential oil blend. In addition,

combining oils that share the same constituents reinforces their effectiveness and offers support from the trace components.

Similarly, when considering the potential contra-actions of using an essential oil, it is necessary to consider its ingredients as part of the overall blend. Scientific tests have shown that skin-sensitizing aldehydes can be neutralized when balanced within the blend of other chemicals in the essential oil. It's important to remember, too, that when essential oils are combined, a new formula is created, often enhancing the effectiveness of the individual oils, which adds to the complexity of their chemical makeup.

Classification of Essential Oils by Chemical Groups

Essential oil compounds can be categorized into two main chemical groups: hydrocarbons, which include terpenes (monoterpenes, sesquiterpenes, diterpenes, and triterpenes), and oxygenated compounds, which include esters, aldehydes, ketones, alcohols, phenols, and oxides.

The primary chemical functional groups in essential oils are alcohols, aldehydes, alkenes, ketones, esters, ethers, phenols, and phenylpropenes.

Hydrocarbons / Terpenes

When hydrogen and carbon atoms are arranged in a chain formation, they form compounds known as terpenes. This category includes monoterpenes, sesquiterpenes, and diterpenes. Terpenes are known for their detoxifying properties, preventing toxin accumulation in the body

Around 90% of citrus oils contain limonene, a powerful terpene known for its uplifting aroma and strong antioxidant, anti-inflammatory, and antimicrobial properties.

and aiding their elimination. Terpenes formed from acetyl-coenzyme A play a vital role in the body's production of hormones, vitamins, and energy.

Monoterpenes – e.g., pinene

Monoterpenes are composed of 10 carbon atoms and two isoprene units. They help rebalance malfunctioning cellular memory and unify other molecules in an essential oil. Around 90% of citrus oils contain limonene, which can kill viruses. Because they are made up of 10 carbons with a varying number of hydrogens and are relatively small, containing one or more double bonds, they are fairly unstable and prone to oxidation. Despite being potential skin irritants, monoterpenes are generally antibacterial and antiseptic, with stimulating, expectorant, and decongestant properties. They can reprogram miswritten information in cellular memory.

Typically, you will find essential oils rich in monoterpenes as top or middle notes that tend to evaporate quickly. However, because of their size, they do have the ability to penetrate the skin easily. Citrus essential oils such as lemon, orange, and grapefruit contain over 90% monoterpenes. Hence, you see why essential oils high in monoterpenes have a shorter shelf life (one to three years).

Isoprene units are branched five-carbon units. One isoprene unit is called a "hemiterpene," which means "half." Two isoprene units form a monoterpene, which means "one." The chemical components of this family are grouped based on specific similarities in their molecular structure.

Monoterpenes comprise two isoprene units of five carbon atoms, totaling ten with at least one double bond. The names of monoterpenes molecules end in "ene."

Essential oils like lemon are made primarily of monoterpenes, while other essential oils may contain one or more chemical families.

It's important to note that the therapeutic effects of monoterpenes can vary depending on the specific compound and its concentration, as well as how it's used (topically, aromatically, or internally). Additionally, individual responses to essential oils may vary, so it's advisable to consult with a healthcare professional before using them for therapeutic purposes, especially if you have any underlying health conditions or are pregnant or breastfeeding.

Safety Considerations

It is important to store this particular group of oils in a dark, cool place away from heat and oxygen. This is because essential oils high in monoterpenes that have oxidized can cause skin irritation or sensitization. In addition, monoterpenes are not water-soluble and should not be used in the bath unless well-diluted with a carrier oil or base. Limonene, present in all citrus oils, has been researched as a potential monoterpene risk as an oxidized irritant or allergen.

Sesquiterpenes – e.g., bisabolene

Sesquiterpenes comprise three isoprene units with five carbon atoms each, totaling fifteen total carbon atoms. Each one has varying hydrogen atoms depending on how many double bonds there are. Sesquiterpenes are made up of only carbon and hydrogen. The name of sesquiterpene molecules ends in "ene."

Essential oils rich in sesquiterpenes are emotionally grounding and calming and are often characterized by heavier, longer-lasting middle-to-base notes due to their reduced volatility. These compounds can cross the blood-brain barrier, increasing oxygen levels in brain tissue and stimulating the pineal and pituitary glands. Sesquiterpenes possess calming, cooling, antiseptic, antibacterial, analgesic, and anti-inflammatory properties. They are also believed to aid in DNA repair, offering potential benefits for cells affected by cancer. Additionally, sesquiterpenes like caryophyllene and valencene act as liver and gland stimulants, while chamazulene and farnesol exhibit strong anti-inflammatory and antibacterial effects. Farnesol is also noted for its antiviral properties, and chamazulene is found in German chamomile, tansy, and yarrow.

"Sesquiterpenes can cross the blood-brain barrier, increasing oxygen levels in brain tissue and stimulating the pineal and pituitary glands."

The sesquiterpene chemical family is found in certain essential oils; in some cases, over 50% of the oil is sesquiterpenes. Caryophyllene gives a robust, woody scent to many essential oils such as clove, lavender, thyme, black pepper, melissa, cinnamon, and ylang ylang. Some sesquiterpenes are unique to a single oil. Sesquiterpenes also increase the half-life of smaller light molecules, both aromatically and therapeutically. These are called "fixing oils" and are frequently used by perfumers to help make a fragrance last longer. Essential oils such as myrrh and sandalwood both contain over 90% sesquiterpenes.

Safety Considerations

Sesquiterpenes are generally safe and not skin irritants unless oils have oxidized. Essential oils high in sesquiterpenes have a long shelf life, typically six to eight years or longer.

Diterpenes – e.g., camphorene

Diterpenes rarely survive the steam distillation extraction process, so they are usually found in only a few oils in minimal quantities. Clary sage is an example of an essential oil containing diterpenes. With 20 carbon atoms and four isoprene units, diterpenes are antibacterial, antiviral, and antifungal. They can also balance hormones, stimulate the immune system, and effectively rid the body of mucus as expectorants.

Safety Considerations

Diterpenes, while beneficial, can cause skin irritation or sensitization in some individuals; thus, conducting a patch test before broader application is essential. Although diterpenes are less commonly associated with phototoxicity compared to other components like furanocoumarins, it is prudent to avoid applying these oils on skin exposed to sunlight or UV light. Internal use of these oils should be strictly avoided unless under the supervision of a qualified healthcare professional, given the

potential for toxicity. Adhering to recommended dilution guidelines is crucial, as high concentrations of diterpenes can lead to adverse effects. Finally, it is important to be aware of potential interactions with medications, and consulting with a healthcare provider is advisable if any prescription or over-the-counter drugs are being taken.

Alkanes – e.g., nonane, decanal

Alkanes are saturated hydrocarbons characterized by single bonds between carbon atoms. The general formula for alkanes is CnH2n+2. Alkanes are relatively less common in essential oils compared to other hydrocarbons like terpenes. However, some essential oils do contain small amounts of alkanes. For example, basil oil (*Ocimum basilicum*) contains small amounts of alkanes such as n-octane. Cypress oil (*Cupressus sempervirens*) includes small amounts of alkanes like n-heptane. Juniper berry oil (*Juniperus communis*) contains alkanes like n-nonane and n-decane, while lemon oil (*Citrus limon*) includes small amounts of alkanes like n-hexane and n-heptane. Although these alkanes are present only in trace amounts, they contribute to the overall chemical composition and properties of these essential oils. Alkanes help protect plants from harmful organisms and predators and reduce water and mineral loss.

Safety Considerations

Alkanes, being relatively non-reactive hydrocarbons, generally have a lower risk of causing skin irritation or sensitization compared to other essential oil constituents. However, prolonged or excessive exposure can still lead to skin dryness or irritation, so it is important to conduct a patch test before extensive use. Unlike other components, alkanes are not associated with phototoxicity, making them safer for application on skin exposed to sunlight. Despite their relatively benign nature, essential oils with significant alkane content should still be used cautiously during pregnancy and nursing, with professional medical advice recommended. Alkanes in essential oils are considered safe for children and pets when properly diluted. Still, seeking guidance on appropriate dilution ratios is always best to prevent potential adverse effects.

Alkenes – e.g., myrcene, thujene

Alkenes are unsaturated hydrocarbons that contain at least one carbon-carbon double bond, with the general formula CnH2n. Several essential oils contain alkenes, contributing to their characteristic aromas and therapeutic properties. Lemongrass oil (*Cymbopogon citratus*) contains alkenes such as myrcene and citral, with citral being a key component known for its strong lemon scent. Basil oil (*Ocimum basilicum*) includes alkenes like myrcene and ocimene, contributing to its fresh and herbaceous aroma. Bay laurel oil (*Laurus nobilis)* contains 1,8-cineole, an ether with an alkene structure, providing a fresh and camphoraceous scent. Cannabis oil (*Cannabis sativa*) includes various alkenes, including myrcene and β-caryophyllene, contributing to its complex and distinctive aroma. Coriander oil (*Coriandrum sativum*) contains alkenes such as myrcene and limonene, giving it a sweet, spicy, and citrus-like scent. These alkenes are significant in the chemical makeup of these essential oils, influencing their scent profiles and potential therapeutic benefits. Alkenes are abundant in trees and other plants and influence the odor of essential oils. These compounds can be identified by their "-ene" ending. Therapeutic properties include anti-inflammatory, analgesic, relaxing, anticancer, antifungal, and antimicrobial.

Safety Considerations

Alkenes, characterized by their carbon-carbon double bonds, include components such as limonene, pinene, and myrcene that can be prone to oxidation, increasing the risk of skin sensitization and irritation. Therefore, storing these oils properly in dark, airtight containers, away from heat and light, is essential to prevent oxidation. Before using these oils, a patch test is recommended to identify potential skin reactions. Some alkenes, like limonene, can be mildly phototoxic, so it is advisable to avoid applying oils containing significant amounts of these compounds to skin exposed to sunlight or UV light. These oils should also be used with care around children and pets, ensuring proper dilution to minimize adverse effects. Internal consumption of essential oils containing alkenes should only be undertaken with the

guidance of a qualified healthcare professional due to the potential for toxicity.

Additionally, some alkenes can interact with medications, particularly those metabolized by the liver. So, consulting a healthcare provider is necessary when using these essential oils alongside any prescription or over-the-counter drugs. Adhering to these specific safety considerations can effectively manage the risks associated with using essential oils containing alkenes.

Alkynes – e.g., falcarinol

Alkynes are unsaturated hydrocarbons characterized by at least one carbon-carbon triple bond. While alkynes are less common in essential oils than other hydrocarbons, they can still be found in certain essential oils. For example, tagetes oil, derived from the Tagetes plant (known as marigold), contains an alkyne called tagetone. Another example is hyssop oil (*Hyssopus officinalis*), which contains pinocamphone, a compound with an alkyne structure. These alkynes contribute to the unique chemical composition and properties of their respective essential oils, adding to the complexity and potential therapeutic benefits of these oils.

Safety Considerations

Specific safety considerations are necessary when using essential oils containing alkynes, such as those with compounds like α-terpinyl acetate in certain coniferous oils. Alkynes, characterized by their carbon-carbon triple bonds, can be more reactive than other hydrocarbons, potentially

leading to skin irritation or sensitization. A patch test before extensive use is essential to check for adverse skin reactions. Due to their reactivity, alkynes should be used with caution during pregnancy and nursing, and professional medical advice should be sought to ensure safety. Additionally, these oils should be adequately diluted around children and pets to avoid any potential irritation or adverse effects. Internal consumption of essential oils containing alkynes should only be done under the supervision of a qualified healthcare professional due to the potential for toxicity and adverse reactions.

Oxygenated Compounds

This second group of chemical constituents includes compounds such as esters, aldehydes, ketones, alcohols, phenols, and oxides.

Alcohols – e.g., linalool

Alcohols are characterized by the hydroxyl (-OH) functional group attached to a carbon atom. Methanol (CH_3OH) and ethanol (C_2H_5OH) are common alcohols. Alcohols are a significant category in essential oils and can contribute to various therapeutic and aromatic properties. Alcohols are generally safe to use on children or the elderly due to their mild nature. They are usually non-toxic and do not cause skin irritation despite being highly germicidal against bacteria, viruses, and fungi. Alcohols are considered extremely therapeutic in aromatherapy due to their pleasant, uplifting fragrance. They also provide antiviral, antifungal, and antibacterial properties.

Safety Considerations

Specific safety considerations must be considered when using essential oils containing alcohols, such as linalool in lavender oil or menthol in peppermint oil. Alcohols in essential oils are generally known for their antimicrobial and soothing properties, but they can still pose risks if not used correctly. For instance, alcohols like menthol can cause skin irritation or sensitization, especially in individuals with sensitive skin. Conducting a patch test to identify potential adverse reactions before applying these oils more broadly is crucial. Although alcohols are not typically associated with phototoxicity, they can enhance the penetration of other compounds through the skin, potentially leading to increased sensitivity or irritation. During pregnancy and nursing, caution is advised, particularly with oils high in alcohols such as menthol, as they can be more potent and may affect the mother and baby. For children and pets, proper dilution is essential to avoid adverse effects, as their skin can be more sensitive. Internal use of essential oils containing alcohols should be approached with caution and only under the guidance of a qualified healthcare professional due to the potential for toxicity and gastrointestinal irritation.

Furthermore, certain alcohols can interact with medications, particularly those affecting the liver or central nervous system, so it is advisable to consult with a healthcare provider when using these essential oils alongside any prescription or over-the-counter drugs. For instance, menthol in peppermint oil can impact liver enzyme activity, potentially altering the metabolism of various medications. It may also influence the central nervous system, which could interact with CNS depressants

or stimulants. Linalool, found in lavender and coriander oils, similarly affects liver enzymes and could interact with drugs metabolized by the liver. Its sedative properties might amplify the effects of CNS depressants. Borneol, present in rosemary and camphor oils, also has the potential to affect liver enzyme activity and may interact with medications processed by the liver. Additionally, its sedative effects can interact with central nervous system-affecting drugs. Due to these potential interactions, it is advisable to consult with a healthcare provider before using essential oils containing these alcohols, especially if you are on prescription or over-the-counter medications.

Acids – e.g., citric acid, myrtanic acid

Acids are organic compounds characterized by the presence of one or more carboxyl

groups (-COOH), and they play a significant role in the chemical composition of some essential oils. Although not as commonly discussed as other components, acids can be found in certain essential oils and contribute to their therapeutic properties. For example, citric acid, commonly found in citrus oils like lemon and lime, is known for its antimicrobial and astringent properties. Myrtanic acid, found in myrtle essential oil, has shown potential for its anti-inflammatory and antimicrobial effects. Valerenic acid, present in valerian oil, is notable for its calming and sedative properties, making it useful in promoting relaxation and sleep. These acids contribute to the therapeutic qualities of essential oils and can also act as

preservatives due to their ability to inhibit microbial growth.

Safety Considerations

When using essential oils containing acids, safety considerations are crucial due to the potential for skin irritation and sensitization. Acids, being reactive compounds, can cause irritation, especially on sensitive skin or when used in high concentrations. Therefore, proper dilution is essential when applying these oils topically. A patch test is recommended to check for any adverse skin reactions. Additionally, due to their acidic nature, these compounds may interact with certain medications or medical conditions. For instance, those with sensitive skin, open wounds, or dermatological conditions should exercise caution when using oils with high acid content.

Internally, the consumption of essential oils containing acids should be approached with caution and only under the guidance of a qualified healthcare professional. These compounds can affect the body's pH balance and may interact with medications, particularly those affecting the gastrointestinal system or metabolized by the liver. Special care should be taken when using these oils around children, pregnant women, and pets, as their sensitivity may increase the risk of adverse effects. Proper storage of essential oils containing acids is also vital, as exposure to air, heat, and light can lead to degradation and reduced efficacy. Adhering to these safety considerations can help ensure the safe and effective use of essential oils containing acids in aromatherapy and other applications.

Carboxylic Acids – e.g., rosmarinic acid

Organic acids are formed by the oxidation of aldehydes, usually ending in "-ic acid." They are rarely found in essential oils because they are not very volatile. Therapeutic properties include anti-inflammatory and antiallergenic effects. Carboxylic acids have a carboxyl (-COOH) functional group. Acetic acid (CH_3COOH) and formic acid ($HCOOH$) are examples.

Safety Considerations

Carboxylic acids, such as rosmarinic acid found in rosemary and oregano oils, require careful consideration regarding their use in essential oils. Rosmarinic acid is known for its antioxidant, anti-inflammatory, and antimicrobial properties but can also present specific safety concerns. Due to its acidic nature, carboxylic acids can irritate the skin, particularly when used in high concentrations or without adequate dilution. A patch test is recommended to identify potential skin sensitivity before broader application. Additionally, because rosmarinic acid and similar compounds can enhance the effects of other substances, they may increase skin absorption of other components in essential oils, potentially leading to heightened sensitivity or adverse reactions. Pregnant and nursing individuals should use essential oils containing carboxylic acids with caution and seek medical advice due to the potential for systemic effects. These oils should also be adequately diluted around children and pets to minimize any risk of irritation. Internal use of essential oils with

significant carboxylic acid content should only be done under the guidance of a qualified healthcare professional, given the potential for gastrointestinal irritation or toxicity. Furthermore, while carboxylic acids are not commonly known for drug interactions, it is still prudent to consult with a healthcare provider when using these essential oils alongside any prescription or over-the-counter medications.

Monoterpenols – e.g., linalool

Monoterpene alcohols, also known as monoterpenols, are common among the essential oils used by aromatherapists due to their mild yet therapeutic qualities. They are generally antimicrobial, stimulating, warming, strengthening, and diuretic. Monoterpenols are particularly useful in skincare due to their gentle anti-inflammatory properties.

Unlike monoterpene, which is a terpene, monoterpenol is an alcohol. Note the difference in the name ending in "ol." Monoterpenols have a ten-carbon backbone, like monoterpenes, but with a functional group. This specific group (-OH) is called hydroxyl. Monoterpenols usually fall in the middle note category, with many being floral and very aromatic.

Interestingly, the molecule's therapeutic properties will be determined and changed based on the "-OH" location on the molecule. All monoterpenols have this hydroxyl functional group (-OH),

which is how someone can recognize it belongs to the monoterpenol chemical family.

Alcohols are characterized by one or more hydroxyl groups (-OH), the oxygen atom bonded to the hydrogen structure. This is when size matters. If it's a 10-carbon, it's a monoterpenol. If it's a 15-carbon, it's a sesquiterpenol. Their shape resembles a monoterpene, except with a (-OH) attached. Depending on where the functional group attaches, different kinds of monoterpenols will be formed.

Take geraniol, for instance. This primary alcohol is determined when the (-OH) attaches to a carbon atom linked to only one other carbon atom. Secondary alcohols are when two carbon atoms, such as menthol, are connected to the (-OH) carbon. Finally, tertiary alcohols are when three carbon atoms are attached to the (-OH) carbon. Terpinene-4-ol and linalool are examples of tertiary alcohols. These make good antimicrobials.

Many essential oils such as clary sage, lavender, and rosewood contain the chemical component linalool. Research on linalool has found it effective against bacteria (E. Joy Bowles). Linalol is also what makes clary sage and lavender quite sedative. Geraniol is present in palmarosa, geranium, and citronella and is responsible for its antifungal and anti-inflammatory properties. The component responsible for tea tree's antibacterial and antiviral qualities is Terpinen-4-ol, a member of the monoterpenol family. Menthol, which

> *Essential oils rich in sesquiterpenes are beneficial for arthritis due to their ability to inhibit the inflammatory response.*

gives peppermint its cooling effect, which peppermint is notorious for, is also the same chemical molecule responsible for its analgesic and antispasmodic effect. Citronellol can be found in rose, lemon, eucalyptus, geranium, and other essential oils.

Essential oils that contain a significant percentage of monoterpenols are typically very potent with anti-infectious properties such as antibacterial, antiviral, antifungal, and antimicrobial with airborne contaminants. Monoterpenols are mild enough for skincare and great as an antiseptic for healing wounds. Linalol offers a calming, sedative effect on the nervous system.

Safety Considerations

Essential oils containing monoterpenols are non-toxic and mild on the skin and mucous membranes, except for peppermint, which can be irritating. Do not use peppermint for children less than the age of five. In addition, infants should not be allowed to inhale peppermint. Essential oils rich in menthol should be avoided by anyone advised to avoid antimalarials, sulfonamides, chloramphenicol, streptomycin, or aspirin and in cases of cardiac fibrillation.

Essential oils containing geraniol, such as geranium, citronella, rose, and thyme ct geraniol, should be diluted at 5%, as these oils can have mild skin-sensitizing effects. Monoterpenols are considered safe for long-term support for well-being and can help boost one's immune system. Essential oils containing monoterpenols usually have a three to five-year shelf life.

Sesquiterpenols – e.g., bisabolol

Sesquiterpene alcohols, also known as sesquiterpenols, may be antispasmodic, antiphlogistic, anti-inflammatory, and sedative. Oils high in sesquiterpenes are helpful for arthritis as they inhibit the inflammatory response.

Like sesquiterpenes, sequiterpenols have a 15-carbon backbone with added hydroxyl functional group (-OH). Essential oils high in sequiterpenols are considered grounding and tend to be heavier, consisting primarily of base notes.

> *Sesquiterpenols are renowned for their soothing, anti-inflammatory, and antimicrobial properties.*

One distinctive trait of individual plants that contain sesquiterpenols is that they create sesquiterpenols unique to that plant. For instance, patchoulol found in patchouli is rarely found in other oils, as is santalol, which is only in sandalwood. Sandalwood is 85% sesquiterpenols. Components in the sesquiterpenol family end in "-ol."

The sesquiterpenol chemical family is found in most essential oils; in some cases, over 50% of the oil is sesquiterpenol. Caryophyllene gives a strong, woody scent to many essential oils such

Diterpenols are structurally similar to human hormones, potentially allowing them to influence and balance the endocrine system.

as clove, lavender, thyme, black pepper, melissa, cinnamon, and ylang ylang. Some sesquiterpenols are unique to a single oil. Sesquiterpenols also increase the half-life of smaller light molecules, both aromatically and therapeutically. These are called "fixing oils" and are frequently used by perfumers to help make a fragrance last longer. Sesquiterpenol chemical components offer a variety of therapeutic properties depending on the individual essential oil.

Safety Considerations

Sesquiterpenols are known for their soothing, anti-inflammatory, and antimicrobial properties. Due to their relatively low reactivity compared to other essential oil components, sesquiterpenols typically present minimal skin irritation or sensitization risk. However, conducting a patch test before extensive use is still advisable to ensure individual tolerance. During pregnancy and nursing, sesquiterpenol-containing oils are considered safe when used in moderation, but consulting a healthcare provider is always recommended. These oils are also suitable for use around children and pets, provided they are properly diluted to prevent any potential minor irritation. Internal consumption of essential oils with sesquiterpenols should be cautiously approached and supervised by a qualified healthcare professional to avoid adverse effects. The shelf life of essential oils high in sesquiterpenols typically is six to eight years. However, β-eudesmol found in blue cypress is cautioned against anticoagulant drugs, peptic ulcers, or any bleeding disorders. Blue cypress should also be avoided during pregnancy.

Diterpenols – e.g., sclareol

Diterpenols, or diterpene alcohols, are a lesser-known group of compounds in essential oils due to their relatively high molecular weight and boiling point, which makes them less common in steam-distilled products. An example of a diterpenol is sclareol, present in clary sage, though it is found in small quantities that can withstand distillation. Diterpenols are structurally similar to human hormones, potentially allowing them to influence and balance the endocrine system. This similarity can make them helpful in regulating hormonal functions and supporting overall endocrine health. Despite their benefits, the lower concentration and higher weight of diterpenols mean they are less prevalent in essential oils than other terpenes, and their effects are often subtle but significant.

Safety Considerations

Specific safety considerations should be considered when using essential oils containing diterpenols, such as sclareol found in clary sage. Diterpenols, due to their high molecular weight, are less likely to cause skin irritation than more volatile compounds. However, given their potent nature and similarity to human hormones, they can have a pronounced impact on hormonal balance, necessitating caution, especially during pregnancy and nursing. Their strong biological activity means that overuse or high concentrations should be avoided. It is advisable to use essential oils containing

Due to their potency, phenols should be used only in low concentrations for short durations.

diterpenols in moderation and to consult with a healthcare provider if there are concerns about hormonal imbalances or if other medications are being taken. Proper dilution is important for children and pets to prevent potential adverse reactions, though diterpenols are generally considered safe when used appropriately.

Phenols – e.g., thymol

Phenols are potent and should only be used in low concentrations for short periods. They are highly antiseptic, effective at killing and preventing the growth of bacteria, fungi, and viruses, and can stimulate the immune system and aid depression. Due to their strength, phenols can irritate the skin and cause liver toxicity with long-term use. Eugenol can be protective of the liver, but only in low concentrations. In high concentrations, it is hepatotoxic and can cause damage to tissues. Phenols can stimulate the immune system when used in acute illnesses and infections. Phenols are strengthening, analgesic, and anti-inflammatory. Thymol found in thyme is beneficial for acute respiratory infections. Clove bud essential oil is 67% phenols and is excellent for combating infections, but people on blood thinners should avoid it due to its high eugenol content. They contain high levels of oxygenating molecules and have antioxidant properties.

Phenols are known to be reactive as they release their hydrogen hydroxyl (-OH) and bind with other positively charged molecules, including protein, which can be harsh and cause skin damage. Essential oils containing phenols are aggressive and have a strong aroma. The benzene ring (6-carbon ring) distinguishes it from alcohols. The benzene ring is typically drawn with a circle in the middle to signify that all the bonds in the ring are equal. Like alcohols, many phenol names end in "-ol."

Phenols tend to be base to middle notes, lasting longer on the skin. They are also responsible for the fragrance of an oil. The shelf life of essential oils containing phenols tends to be three to five years.

Safety Considerations

Phenols are known for their potent antimicrobial and antiseptic properties but can irritate skin and mucous membranes. Due to their strong reactivity, phenolic compounds can cause significant skin irritation, sensitization, or even chemical burns if not properly diluted. Therefore, essential phenol-rich oils should always be used cautiously in well-diluted formulations. It should not be used on individuals with skin issues or dermatitis. Using only at a 1% dilution with other skin-nourishing oils is recommended to balance your blend. Conducting a patch test before widespread application is essential to identify any potential adverse reactions. Phenols can also be absorbed through the skin, potentially leading to systemic effects. Therefore, their use during pregnancy and nursing should be cautiously approached, and professional advice is recommended. These oils are typically not recommended for children and pets due to their strong irritant properties and potential toxicity. Internal use of phenol-containing essential oils should be avoided unless under strict professional supervision due to the risk of severe gastrointestinal irritation and toxicity. By adhering to these specific safety guidelines, the effective and safe use of essential oils containing phenols can be ensured. Essential oils high in phenols tend to irritate the skin and mucous membranes. Phenol-rich oils can be irritating to the skin in a bath. Essential oils high in eugenol, thymol, or carvacrol

inhibit platelet aggregation. They should not be used by people with blood clotting disorders or individual anticoagulant drugs such as aspirin or warfarin before surgery.

Aldehydes – e.g., citronellal

Known for their citrus-like and sometimes aphrodisiac fragrances, aldehydes are generally calming, anti-inflammatory, and hypotensive. They consist of an oxygen atom double bonded to a carbon atom at the end of the carbon chain and are most effective in lower concentrations. Aldehydes can also be strongly antifungal, antiviral, and antiseptic, although they may cause skin irritation.

Aldehydes are highly reactive and characterized by the group Carbon, Hydrogen, and Oxygen (C-H-O). Traces of many aldehydes are found in essential oils and often contribute to their favorable odors, such as cinnamic aldehyde found in cinnamon and cassia and vanillin aldehyde found in onycha and vanilla. They are generally categorized as anti-infectious and sedative to the central nervous system.

Essential oils such as lemongrass contain over 80% aldehydes containing components neral and geranial. The high citral content causes these oils to irritate the skin, but they are pretty calming when inhaled. They can help reduce stress, lower blood pressure, dilate blood vessels, and act as fever reducers. Citral's antiseptic and antiviral action is effective when applied topically for herpes simplex. Citronellal, with its citrusy lemon scent, can be found in oils such as melissa, lemongrass, lemon, mandarin, eucalyptus lemon, cilantro, and citronella.

Other aldehydes include benzaldehyde, cinnamic aldehyde, cuminic aldehyde, and perillaldehyde. The names of aldehydes end in "-al" or "-aldehyde." You will find hundreds of aldehydes in essential oils. They are primarily responsible for the fragrance in perfumes. Because aldehydes are susceptible to oxidation, their shelf life is usually one to three years.

Safety Considerations

When using essential oils that contain high levels of aldehydes, such as citral, it is crucial to follow specific safety considerations. Aldehydes can be highly irritating to the skin and mucous membranes, and their potential to cause irritation increases if the oils are oxidized. Therefore, using these oils in low dilution—typically not exceeding 1%—is strongly recommended for short-term applications only. Citral, a common aldehyde, can cause significant skin irritation and discomfort. However, blending citral with d-limonene or α-pinene may help mitigate these effects by balancing the overall composition of the oil. Internal use of oils high in aldehydes is not recommended under any circumstances, even in small doses, due to the risk of severe gastrointestinal irritation and toxicity.

Additionally, individuals with glaucoma or estrogen-related cancers should exercise caution, as the high citral content in these oils may interact with medications or exacerbate existing conditions. A healthcare provider is advised to ensure safety and avoid any adverse reactions related to medication interactions. By adhering to these guidelines, the safe and effective use of aldehyde-containing essential oils can be maintained.

> *Oils that contain a significant amount of ketones are beneficial for arresting respiratory infections due to their mucolytic and expectorant properties.*

Ketones – e.g., camphor

Ketones have an oxygen atom double bonded to a carbon atom but always on a carbon atom bonded to two other carbon atoms. They are derived from secondary alcohols. Ketones with closed rings have a typical minty-camphoraceous aroma (from the menthone, camphor, and carvone components). Essential oils high in ketones are considered mucolytic, reducing mucus phlegm. Ketone-rich oils are usually middle notes. Ketone names typically end in "-one," except for camphor, whose secondary name is borneone. Ketones also stimulate cell regeneration and promote tissue formation. They are beneficial for dry asthma, colds, or other upper respiratory system conditions and are primarily found in essential oils such as hyssop, clary sage, and sage. Ketones can be calming and sedative, reduce pain, aid digestion, reduce inflammation, improve scar healing, and regenerate new tissue.

Oils that contain a significant amount of ketones are beneficial for arresting respiratory infections due to their mucolytic and expectorant properties, such as rosemary ct. camphor, spike lavender, and peppermint. Other essential oils rich in ketones are considered analgesic and antispasmodic. Oils such as helichrysum, vetiver, and spike lavender help with scars and healing wounds. Some ketone-rich oils are also analgesic.

Not all ketones carry the same concerns or therapeutic characteristics; therefore, consider each one carefully. Non-toxic ketones include menthone, verbenone, valeranone, vetivone, fenchone, carvone, piperitone, jasmone, and turmerone. Additional ketone chemical components should be used at very low dilution and with extreme caution. Those ketones include camphor, isopinocamphone, pinocamphone, pulegone, and thujone.

Hyssop essential oil is high in isopinocamphone and pinocamphone and is considered neurotoxic and convulsant for people with epilepsy. It is also contraindicated during pregnancy. In these cases, limit use to topically only and do not take internally.

Not all essential oils that contain ketones are toxic. Some popular essential oils such as rosemary, eucalyptus, and helichrysum contain a moderate amount of ketones but can still be of therapeutic benefit when moderated in a blend.

The shelf life of essential oils high in ketones is typically three to five years, which is relatively short compared to other essential oils.

Safety Considerations

Ketones, due to their chemical properties, can accumulate in the liver and pose a risk of hepatotoxicity with prolonged or high-dose use. This accumulation occurs because ketones can resist metabolism by the liver, leading to potential liver damage over time. Therefore, limiting ketone-rich oils to a 1% dilution or lower is crucial to minimize the risk of adverse effects. Additionally, ketones can overly stimulate the nervous system, so their use should be moderated.

Pregnant individuals should avoid essential oils high in ketones, as these compounds may harm both mother and fetus.

If ingested by children, camphor can be fatal (4 teaspoons). Camphor can cause convulsions from oral ingestion or inhalation (although rare). Camphor is also known to cause eye and mucous membrane irritation. Studies have linked camphor with miscarriage only in fatal doses. Although there is ample proof that using camphor in moderate amounts is not fetotoxic or teratogenic, it is recommended to avoid essential oils such as ravintsara ct. camphor (84%), feverfew (44%), and Spanish lavender (56%) during pregnancy. Camphor is considered a CNS stimulant, although overdose can cause CNS depressants.

Pregnant individuals should avoid essential oils with high ketone content, as these compounds may have adverse effects on both the mother and the developing fetus. Essential oils such as pennyroyal, calamint, and buchu are high in pulegone, considered hepatotoxic, and contraindicated during pregnancy. Pulegone is also found in small amounts in peppermint, cornmint, and palo santo. Thujone considered neurotoxic and hepatotoxic, is found in mugwort, sage, thuja, western red cedar, wormwood, and in small amounts of blue yarrow and hyssop. It is contraindicated during pregnancy and should not be taken internally. Essential oils high in ketones can potentially irritate the nervous system or cause neurotoxicity in high doses, so individuals with a seizure disorder such as epilepsy should avoid these. In particular, thuja and wormwood oils are rarely used in aromatherapy due to their high ketone content. It is important to adhere to the recommended dosage of essential oils that contain ketones, as they are potentially harmful substances that can accumulate in the body. It is advised to use such oils for a short-term period to avoid possible neurotoxicity.

Esters – e.g., linalyl acetate

Esters are chemical compounds formed when alcohols react with acids. While most plant acids are water-soluble and not abundant in steam-distilled essential oils, esters are more commonly found in hydrosols. Even in minute amounts, esters play a key role in the fruity note of essential oils' fragrance and provide antifungal, anti-inflammatory, sedative, and calming effects on the skin and nervous system.

Esters are formed when an esterification reaction occurs between an alcohol and a carboxylic acid functional group. An ester is essentially a combination of these two functional groups. It includes a carbon atom with a double bond to an oxygen molecule, a single bond to an oxygen molecule, and a single bond to another carbon molecule (usually written -COO- or -COOC-).

Esters can be identified from the proper chemical name, which usually ends with the suffix "-yl, " "-ate, " or "-ester." Common esters in essential oils are methyl salicylate, linalyl acetate, and neryl acetate. Esters are formed from alcohols and acids through a reaction and are named after

both their original molecules, with the alcohols dropping the "ol" and gaining an "yl" and the acids dropping the "ic" and gaining an "ate." For example, linalool (alcohol) and acetic acid (an acid) combine to become "linalyl acetate." Esters have an oxygen double bonded to a carbon bonded to another oxygen.

Esters are calming and soothing. Their aromas tend to be floral and fruity. They tend to be middle notes, some closer to the top than others.

Esters are very common and are found in many essential oils. They are antispasmodic,

anti-inflammatory, antifungal, calming, and relaxing. Linalyl acetate is found in bergamot, clary sage, petitgrain, and lavender. Benzyl benzoate is found in ylang ylang and offers sedative, antispasmodic, and anti-inflammatory qualities, while geranyl acetate is located in geranium and is responsible for its antispasmodic properties. Esters are known for their gentle action and fragrant or fruity notes. Some oils like cardamom, Roman chamomile, and bergamot are good digestive aids. Ester-rich oils are generally effective on skin rashes and skin irritations. Several are adaptogenic, assisting the body in handling stress. Their shelf life is typically three to five years.

Safety Considerations

Esters are generally considered safe for topical application when properly diluted, but there are notable exceptions that require careful attention. One concern involves methyl salicylate in essential oils such as birch and wintergreen. Methyl salicylate can become toxic when absorbed through the skin in significant amounts, especially with repeated application or on damaged skin. Symptoms of poisoning may include dizziness, nausea, vomiting, convulsions, fever, rapid and labored breathing, mental confusion, and kidney failure. Consequently, oils containing methyl salicylate are contraindicated for individuals with asthma, pregnant or breastfeeding women, those with aspirin sensitivity clotting disorders, as well as children, individuals with ADD/ADHD, skin conditions, or a weak constitution.

Another exception is sabinyl acetate, found in *Juniperus sabina*, which is associated with liver toxicity. Essential oils containing sabinyl acetate should be avoided during pregnancy and breastfeeding due to their potential adverse effects. By adhering to these safety considerations, esters in essential oils can be managed more effectively, ensuring safe application and minimizing potential risks.

Oxides – e.g., cineol

Oxides aid the respiratory system due to their expectorant properties. A common example is cineol, the main component of eucalyptus oil. Oxides may also have diuretic, antiseptic, and immune-stimulating actions. Oxides are molecules where an oxygen atom is included in the structure to form a ring, with 1,8-cineole (eucalyptol) being the most common.

Oxides are not a functional group but a group of compounds. Like ethers, oxides have an oxygen atom in the carbon chain, but it is usually combined in the shape of a ring. They are typically derived from alcohols with the -OH group modified as the hydrogen is removed and the oxygen bonds to another carbon to form a ring.

The most common oxide is 1,8-cineole in the chemical component family. Most essential oils contain 1,8-cineole, except for German chamomile. Wikipedia states that 1,8-cineole is eucalyptol and is also known by various synonyms: 1,8-cineole, cajeputol, 1,8-epoxy-p-menthane, 1,8-oxido-p-menthane, eucalyptol, eucalyptol, 1,3,3-trimethyl-2-oxabicyclo[2,2,2]octane, cineol, cineole.

Research shows that 1,8-cineole can be beneficial for respiratory disorders through its anti-inflammatory and mucolytic properties.

Oxides such as 1,8-cineole are a powerful respiratory aid and can have antiviral, antifungal, antibacterial, expectorant, and mucolytic effects. Essential oils containing oxides can act as expectorants as they stimulate mucus secretion and activate the cilia of the mucous membranes. When inhaled, these oils stimulate the brain and increase blood flow. Oils with 1,8-cineole can decrease shortness of breath and have an anti-inflammatory effect on bronchial asthma (see safety considerations).

The shelf life of essential oils containing oxides is two to three years.

Safety Considerations

It is essential to consider the specific oil, dilution, application method, and circumstances of use. People with asthma must exercise caution when using essential oils containing 1,8-cineole, as it may potentially trigger an asthma attack. It is crucial to consider the specific oil, dilution, application method, and circumstances of use. Despite this caution, research indicates that 1,8-cineole can be beneficial for respiratory disorders due to its anti-inflammatory effects on the airways, potentially reducing the need for oral steroids. However, essential oils high in 1,8-cineole can oxidize over time, leading to skin irritation and sensitization. It is recommended to avoid using these oils on infants or children under ten years old. Additionally, oils containing 1,8-cineole may counteract homeopathic remedies and should not be used with animals. By adhering to these safety considerations, the risks associated with 1,8-cineole in essential oils can be effectively managed while taking advantage of its therapeutic benefits.

Ethers – e.g., estragole

Ethers are antispasmodic and carminative in nature. As molecules, they result when oxygen is positioned between two unconnected carbon chains. Essential oils high in ethers are typically avoided and rarely used in aromatherapy due to toxicity concerns. Ethers tend to have strong notes, such as anise's licorice-like aroma. Ethers do not occur in essential oils as simply ethers. Many ethers in essential oils are phenolic, derived from the hydroxyl group of the phenol. This happens when hydrogen from the hydroxyl group breaks off and is replaced with a short hydrocarbon chain (usually a methyl group such as -CH3). Because of the stability of the benzene ring and the methyl group (-CH3), they are considered effective in reducing pain and muscle spasms. Phenolic ethers are notarized as being strong in odor and strong in action. Ether names end in "-ol" or "-ole."

Phenolic ethers are the most broadly found ethers in essential oils. Anethol is found in aniseed and methyl chavicol in basil and tarragon. Ethers that raise toxicity concerns are trans-anethole, methyl eugenol, methyl chavicol, safrole, apiole, and myristicin. Avoid using oils with these constituents during pregnancy. Oils high in ethers should never be used long-term. It's best to use them only for a few days.

It's important to note that while ethers contribute to these therapeutic effects, essential oils contain a complex mixture of compounds, and their overall therapeutic properties depend on the specific combination of constituents present. Additionally, essential oils should be used cautiously and preferably under the guidance of a qualified aromatherapist or healthcare professional, as they can be potent. They may cause adverse reactions in some individuals.

Essential oils with ethers have been described as esters on steroids. The chemical component ether is considered antispasmodic, carminative, analgesic (for the colon, the urinary tract, and

the reproductive system), and effective against some microorganisms. Ethers are stable oils with a shelf life of three to five years.

Safety Considerations

When using essential oils containing ethers, there are several safety concerns to be aware of. Ethers can be hepatotoxic, causing toxicity to the liver, which can lead to liver damage with prolonged or excessive use. They are also genotoxic, meaning they can interfere with the structure and function of DNA, potentially causing abnormal cell changes that could indicate possible, but not definite, carcinogenesis. Neurotoxicity is another concern, as ethers can have toxic or destructive effects on nervous tissue, resulting in symptoms such as dizziness, headaches, nausea, blurry vision, seizures, and coordination problems. Additionally, some ethers have abortifacient properties, meaning they can cause the abortion of a fetus, making them particularly dangerous during pregnancy. Ethers can also exhibit psychotropic effects, influencing the brain and altering mood and behavior.

Furthermore, certain ethers possess estrogen-like activity, behaving like estrogen in the body and potentially disrupting hormonal balance. Due to these significant risks, it is essential to use essential oils containing ethers with extreme caution and awareness of these potential adverse effects. For example, safrole, when used orally in high doses, was shown to be genotoxic, toxic to the kidneys and liver, and is considered to be potentially carcinogenic. Dilutions at 0.01% are considered safe when using essential oils on the skin.

Essential oils high in ethers are rarely used in aromatherapy due to toxicity concerns. Ethers tend to have strong notes, such as anise's licorice-like aroma. Ethers do not occur in essential oils as simply ethers. Many ethers in essential oils are phenolic, derived from the hydroxyl group of the phenol. This happens when hydrogen from the hydroxyl group breaks off and is replaced with a short hydrocarbon chain (usually a methyl group such as -CH3). Because of the stability of the benzene ring and the methyl group (-CH3), they are considered effective in reducing pain and muscle spasms. Phenolic ethers are notarized as being strong in odor and strong in action. Essential oils with ethers have been described as esters on steroids. The

chemical component ether is considered antispasmodic, carminative, analgesic (for the colon, the urinary tract, and the reproductive system), and effective against some microorganisms.

Phenolic ethers are the most broadly found ethers in essential oils. Anethol is found in aniseed and methyl chavicol in basil and tarragon. Ethers that raise toxicity concerns are trans-anethole, methyl eugenol, methyl chavicol, safrole, apiole, and myristicin. Avoid using oils with these constituents during pregnancy. Oils high in ethers should never be used long-term.

Safety Considerations

Ethers in essential oils have several concerns when used in high doses. Safrole, when used orally in high doses, was shown to be genotoxic, toxic to the kidneys and liver, and is considered to be potentially carcinogenic. Dilutions at 0.01% are considered safe when using essential oils on the skin.

The chemical component apiole, present in parsley leaf and parsley seed essential oil, is contraindicated during pregnancy—with a high risk of an abortion if used orally. The oral dose is considered poisonous and can be toxic to the liver, kidneys, heart, and digestive system. External use is not advised either.

Methyl eugenol in high doses is considered genotoxic and carcinogenic. It can also be skin sensitizing, so dilute properly before using topically. Essential oils with small amounts of methyl eugenol include laurel leaf, rose, tarragon, nutmeg, and myrtle.

Myristicin, found in nutmeg, parsnip, parsley leaf, and parsley seed essential oils, can be psychotropic when taken orally. However, myristicin in dried or fresh spices is unlikely to pose any human health risk. It is poisonous to cats. Myristicin is considered low-risk genotoxicity.

Estragole or Methyl chavicol in high doses is genotoxic, carcinogenic, and hepatotoxic in rats and possibly humans. This is contraindicated for anticoagulant medications and any bleeding disorders. When used in aromatherapy, it should be in very low doses. Estragole is found in tarragon, basil ct. estragole, and ponderosa pine.

The chemical component trans-anethole has weak estrogen-like activity and should, therefore, be avoided during pregnancy and breastfeeding or if you have endometriosis or any estrogen-related cancers. It is also cautioned not to be used in oral doses in conjunction with anticoagulant medication, major surgery, peptic ulcer, or another bleeding disorder. Trans-anethole can be found in a high percentage of anise and fennel essential oils.

Lactones – e.g., isoalantolactone

Lactones are an ester group derived from lactic acid. Although lactones can be neurotoxic, essential oils usually contain extremely low quantities. They are primarily known for their balancing and decongestant properties, making them suitable for treating chronic bronchitis. They may cause photosensitivity, such as bergapten found in bergamot. Lactones are mild in aroma but generally strong in action. Lactones and ketones are similar in action since they have some neurotoxic effects and can cause skin sensitization. However, sesquiterpene lactone called helenalin, found in arnica oil, is responsible for its anti-inflammatory action. Trace amounts of lactones are found in German and Roman chamomile, yarrow, and laurel.

Sesquiterpene lactone called helenalin, found in arnica oil, is responsible for its anti-inflammatory action.

Lactones are recognized by their common names, usually ending in "– lactone" but may also end in "-in," "-ine," "-ene," or "-one."

Safety Considerations

When using essential oils containing lactones, such as those found in arnica and elecampane, it is important to be mindful of specific safety considerations. Lactones can be highly sensitizing to the skin and may cause allergic reactions, especially in individuals with sensitive skin or pre-existing skin conditions. Proper dilution is crucial to minimize the risk of irritation and sensitization. Additionally, lactones can be phototoxic, which can cause skin reactions when exposed to sunlight, so it is advisable to avoid sun exposure after applying oils containing lactones. These compounds can also be hepatotoxic, potentially causing liver damage with prolonged or excessive use. Therefore, oils rich in lactones should be used cautiously and not for extended periods. Pregnant and breastfeeding women should avoid using essential oils containing lactones due to the potential risk to the mother and child. Internal use of these oils is generally not recommended unless under the strict supervision of a qualified healthcare professional, as lactones can cause gastrointestinal irritation and other adverse effects.

Furanoids – e.g. bergapten

Furanoids, also known as furanocoumarins, are a distinctive group of oxygenated compounds characterized by a furan ring fused to a coumarin structure. These compounds are primarily found in certain plants and essential oils, notably in citrus oils like bergamot, grapefruit, and lime, as well as in other botanical sources such as figs, parsley, and celery. Furanoids are renowned for their phototoxic properties, meaning they can cause skin irritation or sensitivity when exposed to sunlight. This phototoxicity is due to their ability to absorb ultraviolet light, leading to reactions that can damage skin cells. Despite their potential for causing skin reactions, furanoids are also used therapeutically, particularly in the treatment of skin disorders such as psoriasis and vitiligo, where compounds like psoralen and xanthotoxin are applied in controlled phototherapy treatments. Overall, furanoids play a complex role in the beneficial and adverse effects of certain essential oils and plant extracts.

Furanoids contribute to the reassuring and uplifting effects of citrus oils, adding to their therapeutic value. Coumarins, a lactone found in some essential oils, share similar properties and offer additional benefits. They possess mucus-moving and expectorant properties, which make oils like elecampane effective in treating bronchitis and chest complaints. Moreover, coumarins have demonstrated biological activity for several medical uses, including pharmaceuticals for treating asthma and lymphedema. This dual nature of furanoids and coumarins highlights their complex role in both the therapeutic and safety considerations associated with essential oils and plant extracts.

Safety Considerations

Furanocoumarins can be phototoxic, causing burning to skin exposed to the sun within a few hours after use. Bergapten and bergamottin, present in bergamot, can have this effect, even when used in small amounts in a blend. The lactones and coumarins generally found in essential oils are very low and should not be problematic. However, knowing the specific concentrations in the oils being used is crucial. For example, steam-distilled lime and sweet orange essential oils are not phototoxic

and are safer alternatives for topical use. When using essential oils that contain furanocoumarins, it is recommended to adhere to the maximum safety levels suggested for topical application to minimize the risk of phototoxic reactions. This means avoiding sun exposure or tanning beds for at least 12-24 hours after applying these oils to the skin. For bergamot oil, the concentration should not exceed 0.4% to avoid phototoxicity. By following these guidelines, the risks associated with furanocoumarins can be effectively managed, allowing for their safe and beneficial use in aromatherapy and skincare.

Follow safety guidelines to avoid phototoxicity when using oils with furanocoumarins—limit concentrations and avoid sun exposure.

Furanocoumarins, compounds present in essential oils like bergamot, can lead to phototoxic reactions when the skin is exposed to sunlight after application. This can result in skin irritation or burns. However, non-phototoxic alternatives such as steam-distilled lime and sweet orange offer safer options for those concerned about such reactions. By opting for these alternatives, users can still enjoy the uplifting and aromatic benefits of citrus scents without the risk of phototoxicity.

Furanocoumarins can cause phototoxicity, leading to skin burns when exposed to sunlight. Bergapten and bergamottin in bergamot can trigger this reaction, even in small amounts within blends.

Understanding essential oil chemistry is crucial for crafting effective aromatherapy blends, guiding whether a treatment should be potent and short-term or gentle and extended.

CHAPTER FOUR

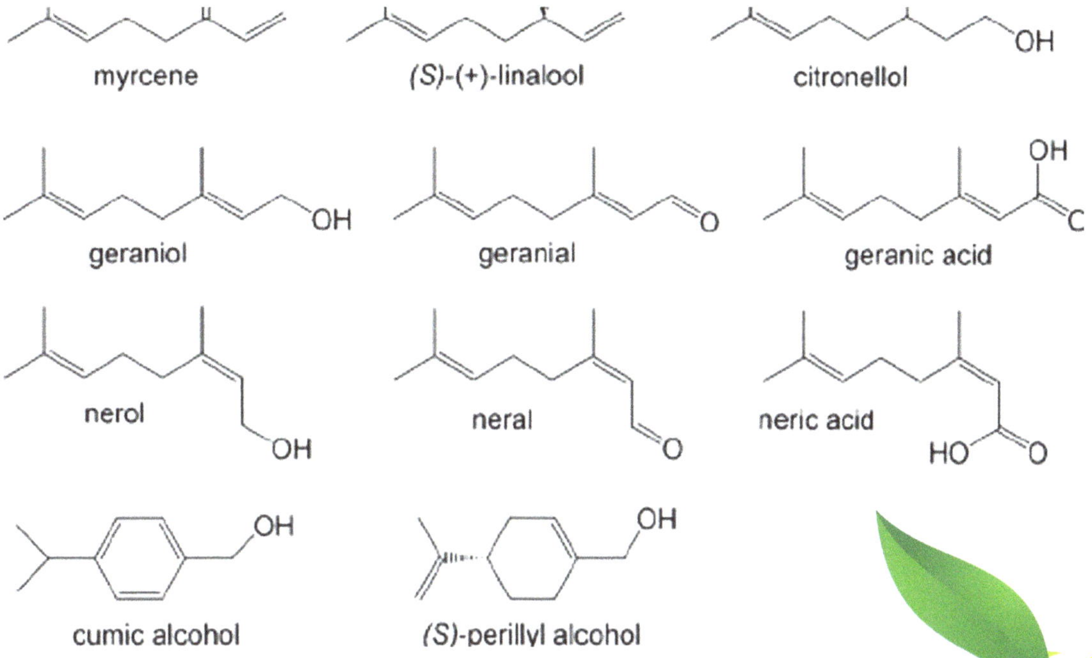

Chemical Component Summary

This summary delves into seven of the most prevalent components found in essential oils. Through this detailed examination, readers will gain valuable insights into the diverse compounds that contribute to the therapeutic efficacy of essential oils, providing a solid foundation for further exploration and mastery of aromatherapy.

1,8-Cineole

1,8-Cineole is in the oxide chemical family. You might be able to tell if an oil has this ingredient by the way it smells. The 1,8-cineole scent opens up the respiratory system. It is the main ingredient in eucalyptus oil and gives it its fresh, clean scent. It also has the name "eucalyptol," which makes it smell so good. 1,8-Cineole, is a versatile compound with numerous therapeutic benefits. It is widely recognized for its potent anti-inflammatory, analgesic, and antimicrobial properties, making it valuable in treating respiratory conditions like asthma, sinusitis, and bronchitis. 1,8-Cineole is an effective mucolytic and expectorant, breaking down mucus and facilitating expulsion. It also serves as a decongestant, helping relieve nasal and sinus congestion. Its cognitive benefits, including enhancing memory and concentration, add to its therapeutic appeal. 1,8-Cineole also exhibits antioxidant properties, protecting cells from damage caused by free radicals. Moreover, it can act as a natural insect repellent, helping to keep insects at bay, and provides a calming effect when inhaled, reducing stress and anxiety.

Benefits of 1,8-Cineole (Eucalyptol)

1,8-Cineole, also called eucalyptol, is a versatile compound renowned for its soothing aroma and therapeutic effects. With powerful anti-inflammatory, analgesic, and antimicrobial properties, it is particularly effective in treating respiratory conditions like asthma, sinusitis, and bronchitis by breaking down mucus and aiding in its expulsion.

Additionally, 1,8-cineole exhibits antiviral properties, supports mental clarity, and can ease muscle and joint discomfort, making it a valuable component in both respiratory and topical formulations.

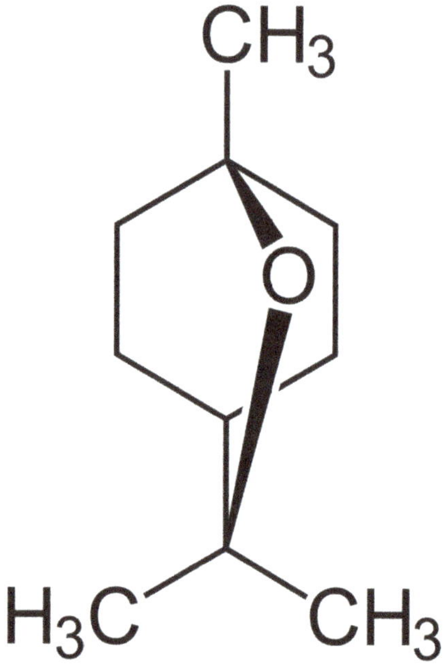

1,8-Cineole boosts cognitive performance by enhancing memory and concentration through increased cerebral blood flow.

Based on scientific research, the benefits and properties of 1,8-cineole include:

- **Anti-inflammatory**: 1,8-Cineole reduces inflammation, particularly in respiratory conditions such as asthma and sinusitis, by inhibiting the production of inflammatory cytokines.
- **Analgesic (Pain Relief)**: It has pain-relieving properties and can alleviate minor aches and pains, such as those associated with headaches or muscle pain.
- **Antimicrobial**: It exhibits antibacterial, antifungal, and antiviral properties, making it effective against various pathogens. This is why it's often used in mouthwashes, dental preparations, and cleaning products.
- **Mucolytic and Expectorant**: 1,8-Cineole helps to break down mucus, making it easier to expel, which is beneficial in respiratory conditions such as bronchitis, colds, and sinusitis.
- **Cognitive Performance**: Some studies suggest that 1,8-cineole can enhance cognitive performance, including improvements in memory and concentration, possibly due to its ability to increase cerebral blood flow.
- **Antioxidant**: It has antioxidant properties, helping neutralize harmful free radicals in the body, protecting cells from damage, and reducing the risk of chronic diseases.
- **Decongestant**: Due to its ability to reduce mucus and inflammation, 1,8-cineole can relieve nasal and sinus congestion, making it helpful in treating colds and other respiratory issues.
- **Anti-spasmodic**: It can help to relieve spasms in the respiratory tract, which is helpful for conditions like asthma and other spastic coughs.
- **Aromatherapeutic Benefits**: Inhaling 1,8-cineole can have a calming effect and may help reduce stress and anxiety, contributing to overall relaxation.
- **Insect Repellent**: It can act as a natural insect repellent and is often used in formulations for this purpose, helping to keep insects at bay.

Safety Considerations

Safety considerations for 1,8-cineole-rich oils include potential skin irritation and sensitization, primarily when used in high concentrations or on sensitive skin. It is advisable to perform a patch test before extensive use. Inhalation of high doses can cause respiratory irritation, so it's important to use these oils in well-ventilated areas and avoid excessive inhalation. 1,8-Cineole-rich oils should be used with caution in young children, pregnant women, and individuals with epilepsy, as they may be more susceptible to adverse effects. 1,8-Cineole may be too potent for people with breathing problems like asthma. Have them smell the bottle's lid and then ask if it makes their chest feel open or tight when they touch it. If their chest gets tight or they don't like the oil, don't put 1,8-cineole oils in their blend. It may be too potent for children under five, especially if used near their faces. Children under the age of ten should use it with care.

A few essential oils that contain 1,8-cineole are:

- **Eucalyptus (*Eucalyptus globulus*):** Known for its high 1,8-cineole content, contributing to its robust and refreshing aroma and respiratory benefits.
- **Rosemary (*Rosmarinus officinalis*):** Contains significant levels of 1,8-cineole, which adds to its stimulating and clarifying effects.
- **Tea Tree (*Melaleuca alternifolia*):** Features 1,8-cineole, enhancing its antimicrobial and decongestant properties.
- **Bay Laurel (*Laurus nobilis*):** Rich in 1,8-cineole, making it a popular respiratory and immune support choice.

- **Cajeput (*Melaleuca cajuputi*):** Contains high amounts of 1,8-cineole, contributing to its expectorant and antiseptic qualities.

Citral

Citral is in the aldehyde chemical family. This component is interesting because it's not an individual component. We call "citral" a combination of neral and geranial components. When neral and geranial appear together in significant amounts, they form citral.

> *Citral is a key component found in the essential oils of lemon and lemongrass.*

Citral has been extensively researched and shown to exhibit a range of beneficial effects:

- **Antimicrobial:** Citral possesses strong antimicrobial properties, effectively inhibiting the growth of various bacteria, fungi, and viruses. This makes it useful in formulations aimed at preventing infections and maintaining hygiene.
- **Anti-inflammatory:** Citral has demonstrated significant anti-inflammatory effects. It can help reduce inflammation and alleviate symptoms associated with inflammatory

conditions such as arthritis and other inflammatory disorders.

- **Antioxidant:** Citral exhibits antioxidant properties, which help neutralize free radicals and reduce oxidative stress. This action can contribute to overall cellular health and may help prevent chronic diseases.
- **Anticancer:** Preliminary research suggests that citral may have anticancer properties, including the potential to inhibit the growth of cancer cells and tumors. It may work by inducing apoptosis (programmed cell death) in cancerous cells and disrupting cancer cell proliferation.
- **Digestive Support:** Citral can aid digestion by stimulating digestive enzymes and helping alleviate symptoms of indigestion and nausea. It can also support the overall health of the digestive system.
- **Anti-anxiety:** Research indicates that citral may have anxiolytic (anti-anxiety) effects, helping to reduce anxiety and promote relaxation. It may be beneficial in managing stress and improving mood.
- **Decongestant:** Citral acts as a natural decongestant, helping to relieve nasal and respiratory congestion. It can support respiratory health by clearing mucus and easing breathing.
- **Skin Benefits:** Citral has soothing properties that can reduce irritation and support a healthy complexion. It may also contribute to the treatment of minor skin conditions.
- **Fragrance:** Citral contributes to the fresh, citrusy aroma of essential oils such as lemon and lemongrass, enhancing their use in perfumery, aromatherapy, and cleaning products.

Citral has soothing properties that can reduce irritation and support a healthy complexion.

Citral is a characteristic of the aldehyde chemical family and is a very potent component. It can cause severe irritation.

Safety Considerations

Citral can cause skin irritation or sensitization, mainly when used in high concentrations or undiluted, so diluting these oils properly before topical application is essential. Citral can seriously irritate skin and mucous membranes. Use a low drop count (about a 0.9% dilution) in blends and combine these oils with skin-nourishing carriers and gentle essential oils. Reserve citral-rich oils (and all aldehyde-rich oils) for acute and short-term (rather than long-term, chronic) issues. Conducting a patch test before widespread use can help identify adverse reactions.

Additionally, citral is known for its potential to increase photosensitivity, making it essential to avoid sun exposure after application to prevent skin damage. While citral has antimicrobial and anti-inflammatory properties, excessive use can lead to potential toxicity or gastrointestinal discomfort if ingested. Avoid using citral-rich oils for pregnant women and nursing mothers. Citral-rich oils are too potent for babies and children. Oils rich in citral tend to have short shelf lives because they oxidize quickly, which can cause skin irritation. Keep your citral-rich oils fresh and store them in a closed bottle in a cool, dark area.

A few essential oils rich in citral are:

- **Lemongrass (*Cymbopogon citratus*):** Known for its high citral content, contributing to its fresh, lemony aroma and strong antimicrobial properties.
- **Lemon Myrtle (*Backhousia citriodora*):** Contains significant levels of citral, which adds to its powerful lemon scent and antiseptic effects.

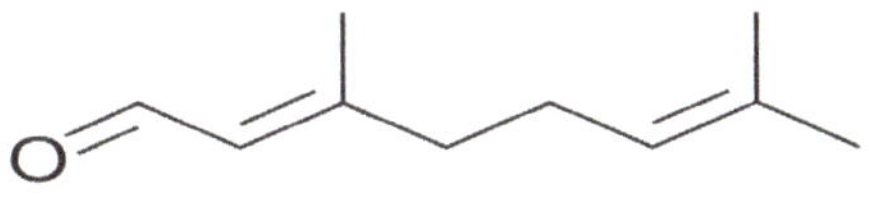

Geranial

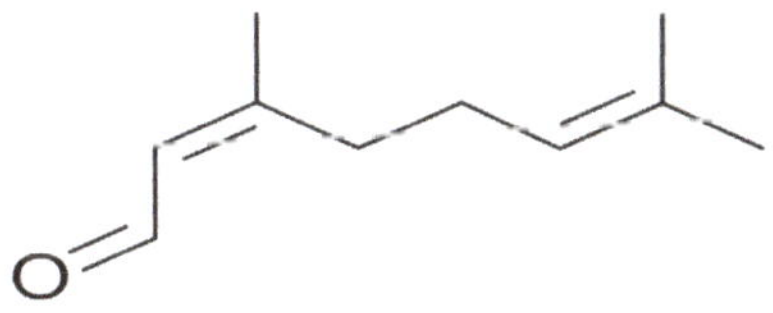

Neral

Citral

$C_{10}H_{16}O$

- **Lemon Tea Tree (*Leptospermum petersonii*):** Rich in citral, making it a popular choice for its uplifting and purifying qualities.
- **May Chang (*Litsea cubeba*):** Features high amounts of citral, enhancing its bright, citrusy fragrance and its use in mood-lifting and cleansing applications.
- **Verbena (*Aloysia citrodora*):** It contains citral, contributing to its sweet, lemony scent and calming and anti-inflammatory properties.

Eugenol

Eugenol is in the phenol chemical family. Clove bud oil has a warm, spicy scent, largely thanks to eugenol. Traces of this aroma show up wherever eugenol goes, alerting you that the oil is incredibly potent against viruses but can be potentially irritating. Oils rich in eugenol are also well known for their warming touch and ability to soothe soreness.

Eugenol has been extensively researched and shown to exhibit several beneficial effects:

- **Antimicrobial:** Eugenol has strong antimicrobial properties, inhibiting the growth of bacteria, fungi, and viruses, making it effective in treating infections.
- **Anti-inflammatory:** It possesses significant anti-inflammatory effects, which can help reduce inflammation and alleviate symptoms of inflammatory conditions.
- **Analgesic:** Eugenol relieves pain, helping alleviate minor aches, toothaches, and other types of discomfort.
- **Antioxidant:** It has antioxidant properties that help neutralize free radicals and reduce oxidative stress, contributing to overall health.

Cinnamon Leaf's rich eugenol content gives it a warm, spicy aroma and potent antiseptic properties.

- **Antiseptic:** Eugenol is known for its antiseptic properties, which help prevent infections in wounds and cuts.
- **Cognitive Benefits:** Some research suggests that eugenol may have potential benefits, including enhancing memory and cognitive function.
- **Muscle Relaxant:** It can act as a muscle relaxant, helping to alleviate muscle tension and spasms.
- **Digestive Support:** Eugenol can relieve digestive health symptoms such as indigestion and nausea.

Safety Considerations

Eugenol can seriously irritate the skin and mucous membranes, so it is advisable to use a low dosage of 0.5% dilution or lower in blends, combining these oils with skin-nourishing carriers and other gentle essential oils. Avoid using eugenol-rich oils on broken or damaged skin, as well as on babies and children. It is also recommended not to use eugenol-rich oils in a bath. These oils and all phenol-rich oils should be reserved for acute issues and short durations rather than long-term or chronic conditions. Additionally, eugenol-rich oils should be avoided if the client is taking MAO-inhibiting or SSRI antidepressant drugs or has liver damage. A patch test is recommended before extensive use to determine sensitivity. Pregnant and breastfeeding women should avoid eugenol-rich oils. Inhalation should be done with caution, especially for those with respiratory issues. If any allergic reactions occur, discontinue use immediately.

A few essential oils rich in eugenol are:

- **Clove (*Syzygium aromaticum*):** Known for its high eugenol content, contributing to its robust, spicy aroma and analgesic properties.
- **Cinnamon Leaf (*Cinnamomum verum*):** Contains significant levels of eugenol, which adds to its warm, spicy scent and antiseptic effects.
- **Basil (*Ocimum basilicum*):** Certain chemotypes of basil, such as the methyl eugenol type, are rich in eugenol, enhancing its antimicrobial and pain-relieving properties.
- **Bay (*Pimenta racemosa*):** Features eugenol, which contributes to its spicy, sweet aroma and is used in muscle and joint pain relief.

Limonene is found in several citrus essential oils, including lemon.

- **Allspice (*Pimenta dioica*):** It contains high amounts of eugenol, giving it its distinctive aroma and beneficial digestive and respiratory health properties.

Limonene

Limonene is in the monoterpene chemical family. It's easy to remember which essential oils contain limonene. The word is related to the Latin name for lemon—*Citrus limon*. Limonene is present in lemon and other citrus essential oils (and a few others).

Limonene has been extensively researched and shown to exhibit several beneficial effects:

- **Antimicrobial:** Limonene possesses antimicrobial properties, helping to inhibit the growth of bacteria and fungi, which can help prevent infections.
- **Anti-inflammatory:** It has anti-inflammatory effects that can help reduce inflammation and alleviate related symptoms in various conditions.
- **Antioxidant:** Limonene exhibits antioxidant properties, neutralizing free radicals and reducing oxidative stress, contributing to overall health and may help prevent chronic diseases.
- **Anxiolytic:** Research suggests that limonene may have anxiolytic effects, helping to reduce anxiety and promote a sense of calm.
- **Digestive Support:** Limonene can support digestive health by aiding in the relief of symptoms such as indigestion and nausea.
- **Cognitive Benefits:** Some studies indicate that limonene may have potential cognitive benefits, including improving mood and cognitive function.
- **Skin Benefits:** Limonene can offer skin benefits, including acting as a natural solvent in formulations and potentially soothing irritated skin.
- **Anticancer:** Preliminary research suggests that limonene may have anticancer properties, potentially inhibiting the growth of cancer cells and tumors.
- **Antiviral:** Limonene has demonstrated antiviral properties that may help combat viruses, including those responsible for colds and flu.

- **Expectorant:** It can act as an expectorant, aiding in relieving congestion by helping to clear mucus from the respiratory tract.
- **Anti-inflammatory:** Limonene's anti-inflammatory effects can help reduce inflammation in the respiratory system, easing symptoms such as sore throat and sinus congestion.
- **Immune Support:** Limonene may support immune function, potentially enhancing the body's ability to fight off infections and reduce the duration of cold and flu symptoms.
- **Decongestant:** It can help relieve nasal and chest congestion, making breathing easier when dealing with respiratory infections.
- **Detoxification Support:** Limonene may support detoxification processes in the body by enhancing liver function and eliminating toxins. Its antioxidant properties also help protect cells from oxidative damage during detoxification.
- **Anti-inflammatory for Muscles and Joints:** Limonene's anti-inflammatory effects can help reduce inflammation and discomfort in muscles and joints, relieving conditions such as arthritis or muscle strains.
- **Analgesic for Muscles and Joints:** It may offer mild pain-relieving effects, helping to alleviate minor aches and pains associated with muscle and joint issues.
- **Improved Circulation:** Limonene can help recover muscles and joints by ensuring nutrients and oxygen reach the affected areas more effectively and promote better circulation.
- **Energetic:** Its pleasant fragrance and diverse uses contribute to limonene's popularity in various industries.

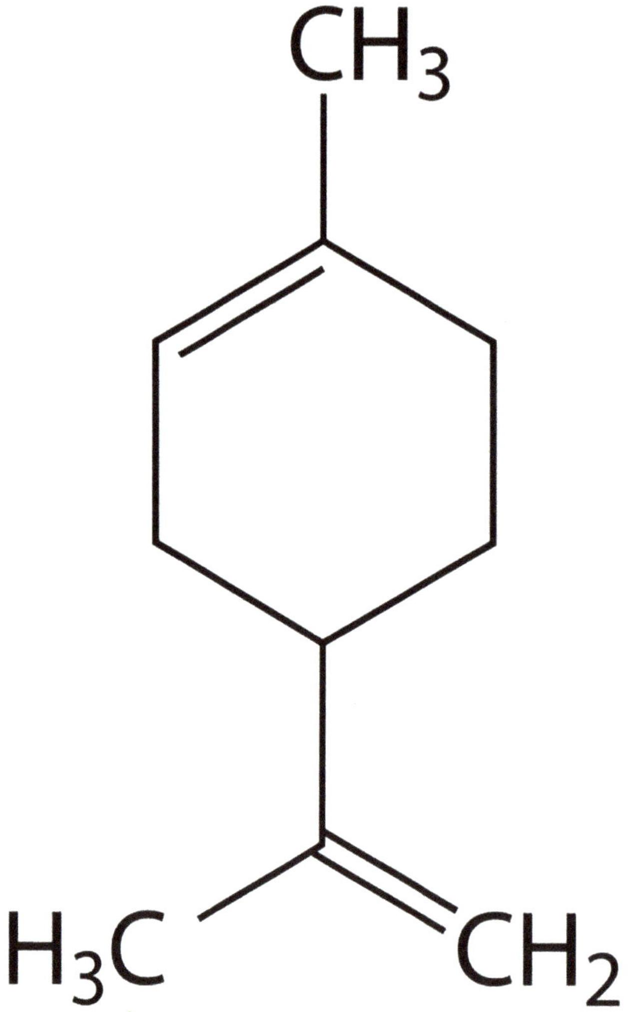

Limonene, a cyclic monoterpene with a six-membered ring and attached methyl and isopropenyl groups, is commonly found in citrus essential oils.

Safety Considerations

Limonene can cause skin irritation and sensitization, especially in high concentrations or undiluted forms. Excessive inhalation of limonene-rich essential oils may lead to respiratory irritation or allergic reactions, particularly in sensitive individuals. Additionally, limonene can increase skin photosensitivity, heightening the risk of sunburn and skin damage if applied before sun exposure. Ingesting limonene-containing products can result in gastrointestinal discomfort or allergic reactions. It is essential to keep limonene oils out of reach of children and pets to prevent accidental ingestion, which may lead to toxicity symptoms such as nausea, vomiting, and central nervous system depression. Pregnant women and individuals with conditions such as asthma or allergies should use limonene-containing essential oils cautiously and seek medical advice before use.

A few essential oils rich in limonene are:

- **Lemon (*Citrus limon*):** Known for its high limonene content, contributing to its fresh, uplifting aroma and cleaning properties.
- **Orange (*Citrus sinensis*):** Contains significant levels of limonene, which adds to its cheerful, energizing scent and mood-enhancing effects.
- **Grapefruit (*Citrus paradisi*):** Rich in limonene, it is popular for its invigorating and refreshing qualities.
- **Lime (*Citrus aurantiifolia*):** Features limonene, which enhances its zesty, revitalizing properties.
- **Tangerine (*Citrus reticulata*):** Contains high amounts of limonene, contributing to its sweet, citrusy aroma and calming effects.

Linalool

Linalool is in the monoterpenol chemical family. Linalool is one of the main components found in lavender essential oil. When this component is in an essential oil, you can make an educated guess that the oil is gentle on the skin and profoundly emotionally calming.

Linalool has been extensively researched and shown to exhibit several beneficial effects:

- **Anti-anxiety:** Linalool has been found to have anxiolytic properties, helping to reduce anxiety and promote relaxation.
- **Sedative:** It has mild sedative effects, which can improve sleep quality and reduce insomnia.
- **Anti-inflammatory:** Linalool can help reduce inflammation, making it useful in alleviating inflammatory conditions.
- **Antimicrobial:** It demonstrates antimicrobial properties, inhibiting the growth of certain bacteria and fungi.
- **Analgesic:** Linalool offers pain-relieving effects, relieving minor aches and discomfort.
- **Antioxidant:** It possesses antioxidant properties that help neutralize free radicals and reduce oxidative stress, contributing to overall health.
- **Skin Benefits:** Linalool can soothe irritated skin and promote a healthy complexion, making it beneficial in topical applications for skin care.
- **Cognitive Benefits:** Some studies suggest that linalool may have potential cognitive benefits, including enhancing memory and reducing cognitive decline.

Linalool and Clary Sage: Relaxation Redefined

Linalool, a key compound in clary sage, supports relaxation and stress relief. Clary sage enhances these benefits with its hormone-balancing properties, making them a perfect pair for tranquility and emotional well-being.

Additionally, clary sage's antispasmodic and sedative effects complement linalool's calming influence, offering relief from tension, headaches, and menstrual discomfort, further promoting both physical and emotional balance.

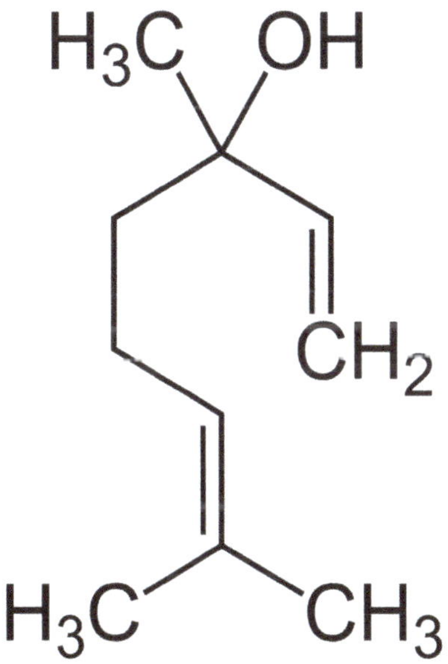

> Linalool has also been shown to interact with certain medications, potentially enhancing their effects.

Safety Considerations

Research indicates that linalool-rich essential oils can cause skin irritation and sensitization. A study published in the *Journal of the American Academy of Dermatology* found that lavender oil, a prominent source of linalool, can lead to allergic contact dermatitis in susceptible individuals (Veien et al., 2004). Inhalation of linalool vapor from essential oils may irritate the respiratory system, particularly those with asthma or other respiratory conditions. A study published in the *Journal of Toxicological Sciences* demonstrated that exposure to linalool vapor can induce airway irritation and inflammation in animal models (Mizoguchi et al., 2012). Individuals may develop allergic reactions to linalool, manifesting as itching, redness, or swelling upon exposure. A review published in *Contact Dermatitis* highlighted cases of allergic contact dermatitis associated with linalool-containing products (Gilpin & Maibach, 2010). Linalool has also been shown to interact with certain medications, potentially enhancing their effects. According to a study published in *Drug Metabolism and Disposition*, linalool inhibits cytochrome P450 enzymes involved in the metabolism of various drugs, leading to altered drug concentrations in the body (Dolara et al., 1996). Additionally, essential oils rich in linalool should be stored securely to prevent accidental ingestion or exposure. The American Association of Poison Control Centers reported numerous cases of accidental ingestion of essential oils by children, resulting in symptoms ranging from mild gastrointestinal upset to more severe toxicity (Gummin et al., 2016).

A few essential oils rich in linalool are:

- **Lavender (*Lavandula angustifolia*):** Known for its high linalool content, contributing to its calming and relaxing properties.
- **Coriander (*Coriandrum sativum*):** Contains significant levels of linalool, which adds to its soothing and digestive benefits.
- **Ho Wood (*Cinnamomum camphora*):** Rich in linalool, making it a popular choice for stress relief and skin care.
- **Basil (*Ocimum basilicum*):** Features linalool, which enhances its anti-inflammatory and calming effects.
- **Rosewood (*Aniba rosaeodora*):** Contains high amounts of linalool, contributing to its uplifting and balancing properties.

α-Pinene

α-pinene is in the monoterpene chemical family. As its name suggests, α-pinene appears in many conifers essential oils (as does its "sibling," β-pinene). Yet this component is present in various other oils, such as frankincense and myrtle.

α-Pinene has been researched and shown to exhibit several beneficial effects:

- **Anti-inflammatory:** α-Pinene can help reduce inflammation, making it helpful in alleviating inflammatory conditions and related symptoms.
- **Antimicrobial:** It has demonstrated antimicrobial properties, inhibiting the growth of bacteria and fungi, which can contribute to its effectiveness in preventing infections.
- **Bronchodilator:** α-Pinene acts as a bronchodilator, helping to open up the airways and

> *α-pinene, a monoterpene, is commonly found in conifers but also appears in frankincense and myrtle, showcasing its diverse presence in essential oils.*

improve breathing, making it beneficial for respiratory conditions like asthma and bronchitis.

- **Cognitive Enhancer:** Research suggests that α-Pinene may enhance cognitive functions, such as memory and concentration, potentially providing neuroprotective benefits.
- **Analgesic:** It exhibits pain-relieving properties, helping to reduce minor aches and discomfort.
- **Antioxidant:** α-Pinene has antioxidant effects that help neutralize free radicals and reduce oxidative stress, which may contribute to overall health.

α-pinene, a monoterpene, is a key component in essential oils from conifers to frankincense and myrtle.

- **Sedative:** It has mild sedative effects that can promote relaxation and reduce anxiety.

Safety Considerations

Essential oils containing α-pinene may cause skin irritation and sensitization, particularly in individuals with sensitive skin. A study published in the *Journal of Toxicology and Environmental Health* demonstrated that

α-pinene can induce skin sensitization reactions (Api et al., 2015). Inhalation of α-pinene vapor from these essential oils may also irritate the respiratory system, leading to symptoms such as coughing or throat irritation. Research published in the *International Journal of Environmental Research and Public Health* reported respiratory irritation and inflammation in individuals exposed to α-pinene vapor (Kim et al., 2019). Direct contact with α-pinene vapor or mist can cause eye irritation and discomfort, as found in a study published in *Regulatory Toxicology and Pharmacology* that observed eye irritation in animal models (Api et al., 2013).

Additionally, some individuals may develop allergic reactions to α-pinene, manifesting as skin rashes, itching, or respiratory symptoms. A review published in *Contact Dermatitis* identified cases of allergic contact dermatitis associated with α-pinene exposure (Frosch et al., 1995). α-Pinene is also highly flammable and should be handled with care to prevent fire hazards, as the National Fire Protection Association classifies it as a flammable liquid with a low flash point (NFPA, n.d.).

A few essential oils rich in α-pinene are:

- **Pine (*Pinus sylvestris*):** Known for its high α-pinene content, contributing to its fresh, woody aroma and therapeutic properties.
- **Rosemary (*Rosmarinus officinalis*):** Contains significant levels of α-pinene, which adds to its stimulating and clarifying effects.
- **Cypress (*Cupressus sempervirens*):** Rich in α-pinene, supporting its use in respiratory and skin treatments.
- **Juniper Berry (*Juniperus communis*):** Features α-pinene, which enhances its detoxifying and purifying properties.

- **Eucalyptus (*Eucalyptus globulus*):** Although primarily known for its 1,8-cineole content, it also contains α-pinene, adding to its respiratory benefits.

Terpinen-4-ol

Terpinen-4-ol is in the monoterpenol chemical family. It's one of the main components found in tea tree essential oil and plays a significant role in its ability to eliminate a wide variety of microbes. Yet, while essential oils rich in this component strongly act against germs, they also tend to be gentle on the skin. Terpinen-4-ol-rich oils are frequently used for an extended time.

Terpinen-4-ol has been shown to:

- **Reduce the presence of germs:** Demonstrates significant antimicrobial activity, helping decrease harmful bacteria and pathogens.
- **Protect immunity during cold and flu season:** Supports the immune system, potentially offering protective benefits during heightened susceptibility to infections.
- **Prevent the growth of fungi:** Effective against various fungal species, aiding in preventing fungal infections.
- **Open breathing passageways during allergy season:** Helps to relieve respiratory symptoms by reducing inflammation and supporting clearer airways.

- **Soothe irritated skin:** Provides calming effects on irritated or inflamed skin, making it helpful in treating minor skin issues.

Safety Considerations

Terpinen-4-ol, a major component of tea tree oil, has the potential to cause skin sensitization and allergic reactions in some individuals. A study published in *Contact Dermatitis* reported cases of allergic contact dermatitis attributed to terpinen-4-ol exposure (Goossens et al., 2009). Direct contact with terpinen-4-ol, especially in its concentrated form, may also irritate mucous membranes, such as the eyes, nose, and mouth. Research published in *Toxicology In Vitro* demonstrated the cytotoxic effects of terpinen-4-ol on human mucosal cells (Hart et al., 2004). Inhalation of terpinen-4-ol vapor or mist can lead to respiratory irritation, with symptoms such as coughing or throat irritation. A study published in *Inhalation Toxicology* found that exposure to terpinen-4-ol vapor resulted in respiratory irritation in animal models (Chan et al., 2018).

Additionally, terpinen-4-ol exhibits estrogenic activity, raising concerns about its potential hormonal effects, particularly in individuals with hormone-sensitive conditions. Research published in *Toxicological Sciences* demonstrated the estrogenic properties of terpinen-4-ol in vitro (Henley et al., 2007). Tea tree oil containing terpinen-4-ol should be stored securely to prevent accidental ingestion or exposure, which can lead to poisoning or adverse reactions. The American Association of Poison Control Centers has reported numerous cases of tea tree oil poisoning in children, with

symptoms ranging from mild gastrointestinal upset to more severe toxicity (Gummin et al., 2016).

Essential oils that contain terpinen-4-ol include:

- **Tea Tree (*Melaleuca alternifolia*):** Known for its high terpinen-4-ol content, contributing to its antimicrobial properties.
- **Cypress (*Cupressus sempervirens*):** Contains terpinen-4-ol, which supports its use in respiratory and skin treatments.
- **Niaouli (*Melaleuca quinquenervia*):** Rich in terpinen-4-ol, adding to its antiseptic and anti-inflammatory effects.
- **Lemon Tea Tree (*Leptospermum petersonii*):** Contains terpinen-4-ol, which enhances its therapeutic benefits.
- **Australian Tea Tree (*Melaleuca armillaris*):** Features terpinen-4-ol, contributing to its effectiveness in treating skin issues.

> *Esters in orange essential oil contribute to its calming effects and sweet, fruity aroma, making it ideal for relaxation and stress relief.*

Esters

Esters are formed when an esterification reaction occurs between an alcohol and a carboxylic acid functional group. An ester is essentially a combination of these two functional groups. It includes a carbon atom with a double bond to an oxygen molecule, a single bond to an oxygen molecule, and a single bond to another carbon molecule (usually written -COO- or -COOC-).

Esters can be identified from the proper chemical name, which usually ends with the suffix "-yl," "-ate," or "-ester." Common esters in essential oils are methyl salicylate, linalyl acetate, and neryl acetate. Esters are formed from alcohols and acids through a reaction and are named after both their original molecules, with the alcohols dropping the "ol" and gaining an "yl" and the acids dropping the "ic" and gaining an "ate." For example, linalool (alcohol) and acetic acid (an acid) combine to become "linalyl acetate." Esters have an oxygen double bonded to a carbon bonded to another oxygen. Esters are calming and soothing. Their aromas tend to be floral and fruity. They tend to be middle notes, some closer to the top than others.

Esters component family are very common and are found in many essential oils. They are antispasmodic, anti-inflammatory, antifungal, calming, and relaxing. Linalyl acetate is found in bergamot, clary sage, petitgrain, and lavender. Benzyl benzoate is found in ylang ylang and offers sedative, antispasmodic, and anti-inflammatory qualities, while geranyl acetate is located in geranium and is responsible for its antispasmodic properties. Esters are known for their gentle action and fragrant or fruity notes. Some oils like cardamom, Roman chamomile, and bergamot are good digestive aids. Ester-rich oils are generally effective in treating skin rashes and skin irritations. Several are adaptogenic, assisting the body in handling stress.

Esters have been shown to be:

- **Antimicrobial:** Esters often possess antimicrobial properties, helping to inhibit the growth of bacteria and fungi.
- **Anti-inflammatory:** They can help reduce inflammation, making them helpful in treating inflammatory conditions.
- **Antioxidant:** Esters may have antioxidant effects, neutralizing free radicals and reducing oxidative stress.
- **Sedative:** Some esters have mild sedative effects, promoting relaxation and reducing anxiety or stress.

Frankincense and Myrrh: Resin and Properties

Frankincense and myrrh, harvested from resinous trees, are rich in terpenes and sesquiterpenes. Frankincense contains boswellic acids, known for their anti-inflammatory properties, while myrrh is valued for its antimicrobial and anti-inflammatory effects. Both resins are integral in aromatherapy for their therapeutic benefits and rich, resinous aromas.

Their combined use enhances their individual effects, creating a powerful synergy for emotional and physical healing.

- **Digestive Support:** Certain esters can aid digestion and soothe gastrointestinal discomfort.
- **Fragrance:** Esters contribute to pleasant and often fruity or floral aromas in essential oils, enhancing their use in perfumery and aromatherapy.
- **Skin Benefits:** They can support skin health by soothing irritation and promoting a healthy complexion.
- **Pain Relieving:** Some esters offer pain-relieving properties, helping to alleviate minor aches and pains.
- **Muscle Relaxant:** Esters can help relax muscles and reduce muscle tension or spasms.

Safety Considerations

Esters are generally safe with proper dilution on the skin, except for two concerns. One is the methyl salicylate, which is present in birch and wintergreen. When this component is absorbed through the skin repeatedly or in large doses or applied to damaged skin, it can be a poison, with symptoms including dizziness, nausea, vomiting, convulsions, fever, rapid and labored breathing, mental confusion, and kidney (renal) failure. Both birch and wintergreen are contraindicated for people with asthma, pregnant or breastfeeding women, and those with aspirin sensitivity and clotting disorders. It should not be used on children, people with ADD/ADHD, skin conditions, or a weak constitution. The second exception is sabinyl acetate, which is present in Juniperus Sabina and is known to cause liver toxicity. Oils containing sabinyl acetate are also contraindicated for pregnancy and breastfeeding.

Essential oils containing esters include:

- **Lavender (*Lavandula angustifolia*):** Contains linalyl acetate, a common ester contributing to its soothing aroma.
- **Bergamot (*Citrus bergamia*):** Rich in linalyl acetate and other esters, giving it a fresh, citrusy scent.

- **Sweet Orange (*Citrus sinensis*):** Includes esters such as ethyl acetate, which enhance its sweet, fruity aroma.
- **Ylang Ylang (*Cananga odorata*):** Contains esters like geranyl acetate, contributing to its exotic fragrance.
- **Rose (*Rosa damascena*):** Contains various esters, including citronellyl acetate, adding to its complex and floral scent.

Lactones

Lactones are an ester group derived from lactic acid. Although lactones can be neurotoxic, essential oils usually contain extremely low quantities. They are primarily known for their balancing and decongestant properties, making them suitable for treating chronic bronchitis. Elecampane is often used for chest complaints. They may cause photosensitivity, such as bergapten found in bergamot. Lactones are mild in aroma but generally strong in action.

As mentioned in the coumarins chemical ester family, lactones and ketones are similar in action since they have some neurotoxic effects and can cause skin sensitization. However, sesquiterpene lactone called helenalin, found in arnica oil, is responsible for its anti-inflammatory action. Trace amounts of lactones are found in German and Roman chamomile, yarrow, and laurel. Lactones are recognized by their common names, usually ending in "–lactone" but may also end in "-in," "-ine," "-ene," or "-one."

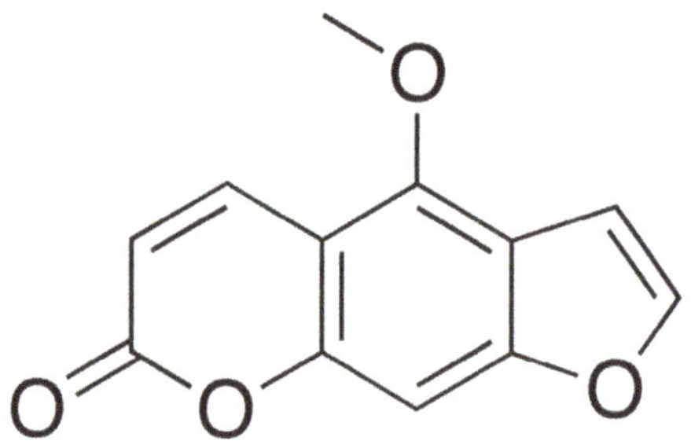

LACTONES

Lactones have been shown to exhibit various properties, including:

- **Antimicrobial:** Inhibits the growth of bacteria, fungi, and other pathogens.
- **Anti-inflammatory:** Reduces inflammation and alleviates symptoms of inflammatory conditions.
- **Antioxidant:** Neutralizes free radicals and reduces oxidative stress.
- **Sedative:** Promotes relaxation and reduces anxiety or stress.
- **Digestive Support:** Stimulates digestive enzymes and soothes gastrointestinal discomfort.
- **Fragrance:** Contributes to pleasant and fruity or floral aromas in essential oils.
- **Skin Benefits:** It supports cell regeneration and repair and is beneficial for maintaining healthy skin.

- **Pain Relieving:** Provides relief from minor aches and pains by influencing pain perception.
- **Muscle Cramps:** Helps relieve muscle cramps by relaxing muscles.
- **Spasms:** Acts as an antispasmodic, reducing involuntary muscle contractions and spasms.
- **Induction of Apoptosis:** Triggers programmed cell death in cancer cells, potentially reducing tumor growth.
- **Inhibition of Tumor Growth:** Slows or stops the proliferation of cancer cells.

- **Anticancer Activity:** Disrupts cancer cell metabolism or signaling pathways essential for tumor survival.
- **Synergistic Effects:** Enhances the efficacy of other cancer treatments when combined with them.

Safety Considerations

While beneficial for their various therapeutic properties, Lactones pose certain safety concerns. Some lactones, particularly those in high concentrations, can be toxic or irritating to the skin and mucous membranes. For instance, compounds such as angelica and costus lactones may cause dermatitis or allergic reactions in sensitive individuals. The potential for liver toxicity is also a concern with prolonged or high-dose use of lactones, necessitating caution, especially for individuals with pre-existing liver conditions. Additionally, certain lactones can interact with medications, potentially altering their effectiveness or causing adverse reactions. Pregnant and breastfeeding women should use lactone-containing products with caution, as their effects on fetal and infant health have not been extensively studied. Overall, it is essential to use lactones within recommended guidelines and to consult healthcare professionals if any adverse reactions occur.

Essential oils that contain lactones include:

- **Clary Sage (*Salvia sclarea*):** Contains sclareolactone, contributing to its characteristic scent.
- **Angelica (*Angelica archangelica*):** Rich in angelica lactones.
- **Costus (*Costus speciosus*):** Contains costunolide and other lactones.
- **Chrysanthemum (*Chrysanthemum cinerariaefolium*):** Includes several types of lactones.

- **Elecampane (*Inula helenium*):** Contains alantolactone and isoalantolactone.

Coumarins

Coumarins are a type of lactone unique because of an adjacent benzene ring with several functional groups attached. Since coumarins are not volatile, they are only found in essential oils by solvent extraction or expression distillation.

The Coumarins component family has been used in treating lymphedema (swelling and puffiness of the lower limbs) and various cancers. Natural coumarins, found in essential oils, are reportedly a stimulant and supporter of the immune system. Coumarins in oils do not act alone but in concert with all the other constituents to restore balance in health and wellness.

Coumarin and its derivatives are all considered phenylpropanoids. Some naturally occurring coumarin derivatives include umbelliferone (7-hydroxycoumarin), aesculetin (6,7 dihydroxy coumarin), herniarin (7-methoxycoumarin), psoralen, and imperatorn.

Coumarins in essential oils exhibit several notable qualities:

- **Anticoagulant:** Coumarins can have blood-thinning properties, which help prevent the formation of blood clots and support cardiovascular health.
- **Antimicrobial:** They possess antimicrobial properties, inhibiting the growth of bacteria and fungi, making them useful in infection treatments.
- **Anti-inflammatory:** Coumarins can reduce inflammation, aiding in the relief of inflammatory conditions such as arthritis and dermatitis.

- **Antioxidant:** These compounds have antioxidant effects, helping neutralize free radicals and reduce oxidative stress, contributing to overall health and longevity.
- **Sedative:** Some coumarins have mild sedative effects, promoting relaxation and potentially helping to alleviate stress and anxiety.
- **Skin Benefits:** Coumarins can support skin health by promoting cell regeneration and repair, which is beneficial for maintaining a healthy complexion and treating minor skin irritations.
- **Fragrance:** Coumarins contribute to the pleasant, sweet, and sometimes spicy aroma of certain essential oils, enhancing their use in aromatherapy and perfumery.
- **Analgesic**: Provides pain relief.
- **Antispasmodic**: Relieves muscle spasms and cramps.
- **Vasodilator**: Dilates blood vessels, improving blood flow and reducing blood pressure.
- **Antitussive**: Suppresses coughing and is beneficial for respiratory issues.
- **Lymphatic Support**: Supports lymphatic drainage and detoxification processes.

Safety Considerations

One of the most notable issues with coumarins, especially those in citrus oils like bergamot, lime, and lemon, is their ability to cause photosensitivity. This reaction can lead to skin irritation and burns when the skin is exposed to sunlight after applying oils containing coumarins. As a result, it's advisable to avoid direct sunlight or tanning beds for at least 12 hours after using such oils on the skin.

Additionally, some coumarins have anticoagulant properties, which can be problematic for individuals on blood-thinning medications or those with bleeding disorders, as they may increase the risk of bleeding. High doses of coumarins can also pose hepatotoxic risks, particularly for those with pre-existing liver conditions,

making it essential to use these oils in moderation and under professional guidance.

Given these potential risks, it is crucial to dilute essential oils containing coumarins properly and conduct a patch test before widespread application. Pregnant or breastfeeding women, children, and individuals with sensitive skin should exercise extra caution or avoid using these oils altogether.

For topical use, it is recommended not to use more than the maximum safety level suggested below if the skin will be exposed to sunlight or tanning beds. Angelica root- 0.8%, Bergamot- 0.4%, Cumin- 0.4%, Grapefruit- 4%, Laurel Leaf absolute- 2%, Lemon- 2%, Lime- 0.7%, Mandarin Leaf- 0.17%, Bitter Orange- 1.25%, Rue- 0.15%, Taget- 0.01%.

Essential oils containing coumarins include:

- **Sweet Clover (*Melilotus officinalis*):** Contains coumarin and its derivatives.
- **Tonka Bean (*Dipteryx odorata*):** Rich in coumarin, giving it a sweet, vanilla-like aroma.
- **Cassia (*Cinnamomum cassia*):** Contains coumarin, contributing to its distinctive spicy scent.
- **Lavender (*Lavandula angustifolia*):** Contains small amounts of coumarin, contributing to its overall profile.
- **Fennel (*Foeniculum vulgare*):** Includes coumarins, adding to its aromatic complexity.
- **Bergamot (*Citrus bergamia*):** Includes Contains bergamottin, bergapten, and umbelliferone, contributing to its mood-lifting effects, Umbelliferone
- **Lemon (*Citrus limon*):** Contains bergapten and limettin, contributing to its refreshing scent.

Furanocoumarins shield the skin from UV radiation, helping to prevent sun-induced damage.

- **Grapefruit (*Citrus paradisi*):** Includes bergapten and auraptene, enhancing its uplifting aroma.
- **Sweet Orange (*Citrus sinensis*):** Contains bergapten and auraptene, which add to its bright, cheerful fragrance.
- **Lime (*Citrus aurantiifolia*):** Contains limettin and bergapten, contributing to its fresh scent.
- **Tangerine (*Citrus reticulata*):** Includes auraptene and bergamottin, enhancing its sweet aroma and mood-lifting effects.
- **Mandarin (*Citrus nobilis*):** Contains auraptene and bergamottin, adding to its calming and uplifting properties.
- **Roman Chamomile (*Chamaemelum nobile*):** Includes umbelliferone and herniarin, contributing to its soothing and anti-inflammatory effects.
- **Angelica Root (*Angelica archangelica*):** Contains imperatorin and xanthotoxin, which add to its grounding and skin-soothing benefits.
- **Lavender (*Lavandula angustifolia*):** Contains trace amounts of coumarin, adding subtle sweetness to its relaxing profile.
- **Clary Sage (*Salvia sclarea*):** Includes umbelliferone and scopoletin, contributing to its calming and hormone-balancing effects.
- **Tonka Bean (*Dipteryx odorata*):** Contains coumarin and umbelliferone, enhancing its rich, sweet aroma.
- **Cassia (*Cinnamomum cassia*):** Contains coumarin, adding warmth and depth to its spicy scent.

Furanocoumarins

Furanocoumarins, also known as furanoids, are a unique class of compounds found predominantly in certain plants and essential oils. Characterized by their distinctive furan ring fused to a coumarin structure, these compounds are notably present in citrus oils like bergamot, grapefruit, and lime and in botanical sources such as figs, parsley, and celery. Furanocoumarins are recognized for their phototoxic properties, which can lead to skin irritation or sensitivity when exposed to sunlight. Despite these potential risks, they are also valued for their therapeutic applications, including treating skin disorders like psoriasis and vitiligo through controlled phototherapy. The dual nature of furanocoumarins—combining

Furanocoumarins, found in mandarins, offer a natural defense against UV radiation, helping to protect your skin from sun-induced damage and maintain its health and vitality.

> Parsley is rich in furanocoumarins, including bergapten.

beneficial uses and safety considerations—highlights the importance of understanding their properties in therapeutic and safety contexts. Furanocoumarins contribute to citrus oils' reassuring and uplifting effects.

Furanocoumarins offer several benefits, including:

- **Photoprotective Effects:** Furanocoumarins can protect against UV radiation, which helps prevent skin damage caused by sun exposure.
- **Antimicrobial Activity:** They possess antimicrobial properties that can help inhibit the growth of certain bacteria and fungi.
- **Anti-inflammatory:** Furanocoumarins can help reduce inflammation, making them useful in treating inflammatory skin conditions and other disorders.
- **Antioxidant Properties:** They have antioxidant effects that help neutralize free radicals, reducing oxidative stress and potentially lowering the risk of chronic diseases.
- **Skin Regeneration:** Furanocoumarins support skin regeneration and healing, making them beneficial for skin repair and maintenance.
- **Immune Support:** They can enhance immune function, helping the body to better defend against infections and diseases.

Safety Considerations

While beneficial for their therapeutic properties, furanocoumarins pose specific safety considerations that must be carefully managed. One of the primary concerns is their phototoxicity, which can lead to skin irritation or sensitivity when exposed to sunlight or ultraviolet light. This phototoxic effect is particularly pronounced in citrus oils such as bergamot, grapefruit, and lime, which contain significant levels of furanocoumarins like bergapten and bergamottin. These compounds can cause severe skin reactions if used in high concentrations or when applied before sun exposure. Additionally, furanocoumarins can interact with certain medications, potentially altering their effectiveness or causing adverse reactions. Pregnant and breastfeeding women should exercise caution, as the effects of furanocoumarins on fetal and infant health are not well understood. To mitigate risks, it is recommended to use essential oils containing furanocoumarins with appropriate precautions, such as avoiding sunlight exposure after application and adhering to recommended dilution guidelines.

Several essential oils contain furanocoumarins, including:

- **Bergamot (*Citrus bergamia*):** Contains significant levels of bergapten and bergamottin.
- **Grapefruit (*Citrus x paradisi*):** Includes various furanocoumarins, contributing to its phototoxic potential.
- **Lime (*Citrus aurantiifolia*):** Contains furanocoumarins such as limettin.
- **Celery (*Apium graveolens*):** Contains furanocoumarins like psoralen.
- **Parsley (*Petroselinum crispum*):** Rich in furanocoumarins, including bergapten.

Ethers

Essential oils high in ethers are typically avoided and rarely used in aromatherapy

> *Some ethers ease muscle spasms and cramps with their antispasmodic effects.*

due to toxicity concerns. Ethers tend to have strong notes, such as anise's licorice-like aroma.

Ethers do not occur in essential oils as simply ethers. Many ethers in essential oils are phenolic, derived from the hydroxyl group of the phenol. This happens when hydrogen from the hydroxyl group breaks off and is replaced with a short hydrocarbon chain (usually a methyl group such as -CH3).

Because of the stability of the benzene ring and the methyl group (-CH3), they are considered effective in reducing pain and muscle spasms. Phenolic ethers are notarized as being strong in odor and strong in action. Ether names end in "-ol" or "-ole."

Phenolic ethers are the most broadly found ethers in essential oils. Anethol is found in aniseed and methyl chavicol in basil and tarragon. Ethers that raise toxicity concerns are trans-anethole, methyl eugenol, methyl chavicol, safrole, apiole, and myristicin. Avoid using oils with these constituents during pregnancy. Oils high in ethers should never be used long-term. It's best to use them only for a few days.

Here are the major properties of ethers:

- **Antimicrobial:** Ethers often possess antimicrobial properties, which can help inhibit the growth of bacteria and fungi.
- **Antispasmodic:** Some ethers have antispasmodic effects, making them helpful in relieving muscle spasms and gastrointestinal cramping.

- **Sedative:** Ethers can have mild sedative properties, helping to promote relaxation and reduce anxiety.
- **Analgesic:** Certain ethers provide pain-relief benefits, assisting in alleviating minor aches and pains.
- **Decongestant:** Ethers can act as decongestants, helping to clear nasal and respiratory congestion.
- **Non-Irritating:** Ethers are generally less irritating to the skin than other essential oil components, making them suitable for sensitive skin applications.
- **Stabilizing:** Ethers can contribute to the stabilization of essential oil blends, improving their shelf life and effectiveness.

Safety Considerations

Ethers in essential oils have several concerns when used in high doses. Safrole, when used orally in high doses, was shown to be genotoxic, toxic to the kidneys and liver, and is considered to be potentially carcinogenic. Dilutions at 0.01% are considered safe when using essential oils on the skin. The chemical component apiole, present in parsley leaf and parsley seed essential oil, is contraindicated during pregnancy—with a high risk of an abortion if used orally. The oral dose is considered poisonous and can be toxic to the liver, kidneys, heart, and digestive system. External use is not advised either. Methyl eugenol in high doses is considered genotoxic and carcinogenic. It can also be skin sensitizing, so dilute properly before using topically. Essential oils with small amounts of methyl eugenol include laurel leaf, rose, tarragon, nutmeg, and myrtle.

Myristicin, found in nutmeg, parsnip, parsley leaf, and parsley seed essential oils, can be psychotropic when taken orally. However, myristicin in dried or fresh spices is unlikely to pose any human health risk. Poisonous to cats. Myristicin is considered low-risk genotoxicity. Estragole or Methyl chavicol in high doses is genotoxic, carcinogenic, and hepatotoxic in rats and possibly humans. This is contraindicated for anticoagulant medications and any bleeding disorders. When used in aromatherapy, it should be in very low doses. Estragole is found in tarragon, basil ct. estragole, and ponderosa pine.

The chemical component trans-anethole has weak estrogen-like activity and should, therefore, be avoided during pregnancy and breastfeeding or if you have endometriosis or any estrogen-related cancers. It is also cautioned not to be used in oral doses in conjunction with anticoagulant medication, major surgery, peptic ulcer, or another bleeding disorder. Trans-anethole can be found in a high percentage of anise and fennel essential oils.

Ethers in essential oils raise several safety concerns due to their potentially harmful effects on the body. They can be hepatotoxic, meaning they may cause liver toxicity and impair liver function. Additionally, ethers are genotoxic, interfering with DNA structure and function, which can lead to abnormal cell changes and a possible risk of carcinogenesis. Neurotoxic effects are also a concern, as ethers may negatively impact nervous tissue, leading to symptoms such as dizziness, headaches, nausea, blurry vision, seizures, and coordination problems. Some ethers are abortifacient, meaning they can induce abortion in pregnant individuals. They may also have psychotropic effects, influencing mood and behavior by affecting brain function. Furthermore, ethers can exhibit estrogen-like activity, mimicking the

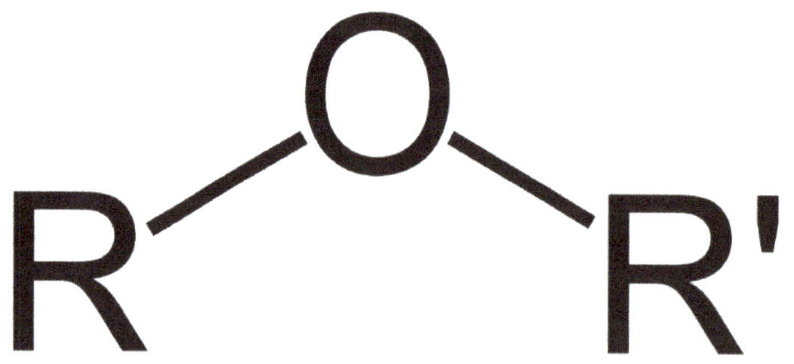

ETHERS

hormone estrogen in the body and potentially affecting hormonal balance.

A few essential oils that contain ethers are:

- **Sweet Fennel (*Foeniculum vulgare*):** Known for its high content of anethole, a key ether component, contributing to its sweet, licorice-like aroma and digestive benefits.
- **Anise (*Pimpinella anisum*):** Rich in anethole, giving it a sweet, licorice scent and properties that support respiratory and digestive health.
- **Tarragon (*Artemisia dracunculus*):** Contains methyl chavicol (estragole), an ether that imparts a sweet, spicy aroma and helps with digestive issues and muscle spasms.
- **Basil (*Ocimum basilicum*):** Some chemotypes of basil are rich in methyl chavicol, which adds to its spicy, herbal fragrance and antispasmodic properties.
- **Star Anise (*Illicium verum*):** Features a high concentration of anethole, contributing to its strong, sweet aroma and uses in supporting respiratory and digestive functions.

Ketones

Ketones have an oxygen atom double bonded to a carbon atom but always on a carbon atom bonded to two other carbon atoms. They are derived from secondary alcohols. Ketones with closed rings have a typical minty-camphoraceous aroma (from the menthone, camphor, and carvone components). Essential oils high in ketones are considered mucolytic, reducing mucus phlegm. Ketone-rich oils are usually middle notes. Ketone names typically end in "-one," except for camphor, whose secondary name is borneone. Ketones also stimulate cell regeneration and promote tissue formation. They are beneficial for dry asthma, colds, or other upper

> *Ketones stimulate cell regeneration, promote tissue formation, and aid in treating dry asthma, colds, and upper respiratory conditions.*

respiratory system conditions and are primarily found in essential oils such as hyssop, clary sage, and sage.

Here are the major properties of ketones:

- **Anti-inflammatory:** Ketones can help reduce inflammation, making them useful in treatments for inflammatory conditions.
- **Antiseptic:** Many ketones have antiseptic properties, which can aid in preventing infections and promoting wound healing.
- **Expectorant:** Some ketones act as expectorants, helping to clear mucus from the respiratory tract and easing coughs and congestion.
- **Analgesic:** Ketones can relieve pain, helping soothe minor aches and pains.
- **Cell Regeneration:** Certain ketones support cell regeneration and tissue repair, which is beneficial for skin health and wound healing.
- **Skin Sensitization:** Ketones can cause skin irritation or sensitization in some individuals, particularly in high concentrations. Proper dilution and patch testing are recommended.
- **Cognitive Support:** Some ketones, like those found in rosemary, may have cognitive benefits, including enhancing memory and concentration.

Safety Considerations

Caution should be exercised when using essential oils rich in ketones, as high daily doses can cause harm to the liver over an extended period of use. This is because ketones can resist being metabolized by the liver, so they tend to accumulate. Be sure to use ketone-rich oils at 1% dilution or less since ketones can be too stimulating for the nervous system. The shelf life of an essential oil high in ketones is usually three to five years. Avoid essential oils high in ketones during pregnancy.

Essential oils high in ketones can potentially irritate the nervous system or cause neurotoxicity in high doses, so individuals with a seizure disorder such as epilepsy should avoid these. In particular, thuja and wormwood oils are rarely used in aromatherapy due to their high ketone content. It is important to adhere to the recommended dosage of essential oils that contain ketones, as they are potentially harmful substances that can accumulate in the body. It is advised to use such oils for a short-term period to avoid possible neurotoxicity.

Not all essential oils that contain ketones are necessarily toxic. Some popular essential oils such as rosemary, eucalyptus, and helichrysum contain a moderate amount of ketones but can still be of therapeutic benefit when moderated in a blend.

Oils that contain a significant amount of ketones are beneficial for arresting respiratory infections due to their mucolytic and expectorant properties, such as rosemary ct. camphor, spike lavender, and peppermint. Other essential oils rich in ketones are considered analgesic and antispasmodic. Oils such as helichrysum, vetiver, and spike lavender help with scars and healing wounds. Some ketone-rich oils are also analgesic.

Not all ketones carry the same concerns or therapeutic characteristics; therefore, consider each one carefully. Non-toxic ketones include menthone, verbenone, valeranone, vetivone, fenchone, carvone, piperitone, jasmone, and turmerone. Additional ketone chemical components should be used at very low dilution and with extreme caution. Those ketones include camphor, isopinocamphone, pinocamphone, pulegone, and thujone.

Hyssop essential oil is high in isopinocamphone and pinocamphone and is considered neurotoxic and convulsant for people with epilepsy. It is also contraindicated during pregnancy. In these cases, limit use to topically only and do not take internally.

Essential oils such as pennyroyal, calamint, and buchu are high in pulegone, considered hepatotoxic, and contraindicated during pregnancy. Pulegone is also found in small amounts in peppermint, cornmint, and palo santo. Thujone considered neurotoxic and hepatotoxic, is found in mugwort, sage, thuja, western red cedar, wormwood, and in small amounts of blue yarrow and hyssop. It is contraindicated during pregnancy and should not be taken internally.

Therapeutic blends made with camphor should never be placed near the noses of young children or infants. If ingested by children, camphor can be fatal (4 teaspoons). Camphor can cause convulsions from oral ingestion or inhalation (although rare). Camphor can cause eye and mucous membrane irritation. Studies have linked camphor with miscarriage only in fatal doses. Although there is ample proof that using camphor in moderate amounts is not fetotoxic or teratogenic, it is recommended to avoid essential oils such as Ravintsara ct. camphor (84%), feverfew (44%), and Spanish lavender (56%) during pregnancy. Camphor is considered a CNS stimulant, although overdose can cause CNS depressants.

A few essential oils that contain ketones are:

- **Pennyroyal (*Mentha pulegium*):** Known for its high content of pulegone, a ketone with strong effects on the liver and nervous system.
- **Spearmint (*Mentha spicata*):** It contains carvone, a ketone that contributes to its fresh, minty aroma and digestive benefits.
- **Carrot Seed (*Daucus carota*):** Rich in β-carotene and other ketones, which support skin health and regeneration.
- **Rosemary (*Rosmarinus officinalis*):** Contains ketones like camphor, which provide stimulating and clarifying effects.
- **Fennel (*Foeniculum vulgare*):** Features ketones such as trans-anethole, contributing to its sweet, licorice-like scent and digestive properties.

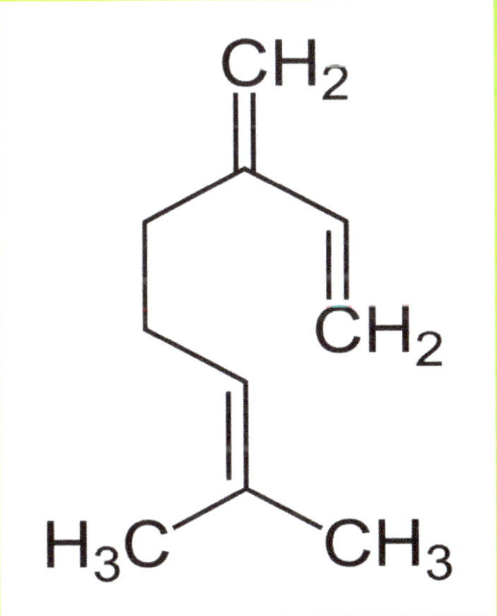

Monoterpenes

In an earlier chapter, terpene was defined as a molecule of carbon and hydrogen only. One of the chemical families with this type of structure is monoterpenes. Because they are made up of 10 carbons with a varying number of hydrogens and are relatively small, containing one or more double bonds, they are fairly unstable and prone to oxidation. Typically, you will find essential oils rich in monoterpenes as top or middle notes that tend to evaporate quickly. However, because of their size, they do have the ability to penetrate

Monoterpenes in Lemon Oil

Monoterpenes found abundantly in lemon oil, are known for their refreshing, uplifting aroma and their natural ability to boost mood and support overall well-being.

They also possess antimicrobial properties that help purify the air and enhance the immune system. Additionally, their invigorating scent can improve mental clarity and focus.

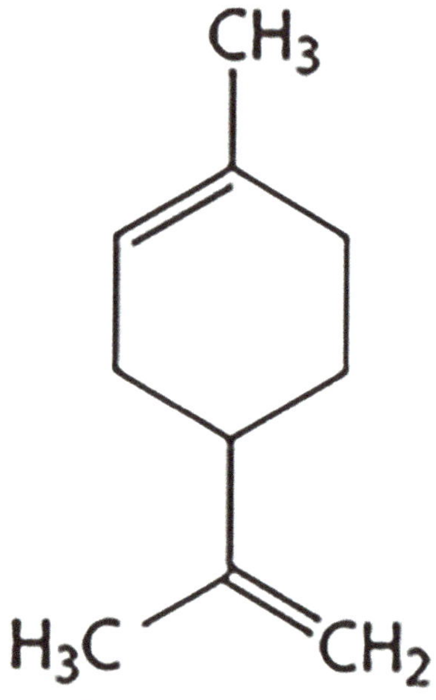

the skin easily. Citrus essential oils such as lemon, orange, and grapefruit contain over 90% monoterpenes. Hence, you see why essential oils high in monoterpenes have a shorter shelf life (one to three years).

Isoprene units are branched five-carbon units. One isoprene unit is called a "hemiterpene," which means "half." Two isoprene units form a monoterpene, which means "one." The chemical components of this family are grouped based on specific similarities in their molecular structure. Monoterpenes comprise two isoprene units of five carbon atoms, totaling ten with at least one double bond. The names of monoterpenes molecules end in "ene." In the monoterpenes component family, essential oils such as lemon are made primarily of monoterpenes, while other essential oils may contain one or more chemical families.

Monoterpenes have been shown to exhibit various properties, including:

- **Antimicrobial Activity:** Many monoterpenes, such as limonene and 1,8-cineole, possess antibacterial, antiviral, and antifungal properties that contribute to their effectiveness in supporting the immune system and protecting against infections.
- **Anti-inflammatory Effects:** Compounds like α-pinene and β-pinene can help reduce inflammation, making them useful in managing inflammatory conditions and promoting overall wellness.
- **Decongestant Properties:** Monoterpenes, particularly 1,8-cineole found in eucalyptus oil, are known for relieving respiratory congestion and improving breathing.
- **Antioxidant Activity:** Some monoterpenes exhibit antioxidant effects, helping to neutralize free radicals and protect cells from oxidative damage.
- **Mood Enhancement:** Monoterpenes such as limonene and menthol can have uplifting and mood-enhancing effects, contributing to their use in aromatherapy for stress relief and emotional balance.

Geranium oil, rich in monoterpenes, promotes emotional balance and well-being.

- **Digestive Support:** Monoterpenes like menthol in peppermint oil can aid digestion by soothing the gastrointestinal tract and alleviating symptoms such as nausea and bloating.

Safety Considerations

Monoterpenes in various essential oils present specific safety considerations that must be carefully managed. These compounds can cause skin irritation and sensitization, particularly when used in high concentrations or undiluted. Essential oils rich in monoterpenes, such as limonene and α-pinene, have been reported to irritate the skin, potentially leading to allergic reactions. Inhalation of monoterpene-rich vapors can also irritate the respiratory system, causing symptoms like coughing or throat irritation, and may exacerbate asthma. Additionally, some monoterpenes, including limonene, can increase photosensitivity, heightening the risk of sunburn and skin damage if applied before sun exposure. It is crucial to store monoterpene-containing oils away from children and pets to avoid accidental ingestion, which can lead to toxicity symptoms such as nausea and gastrointestinal distress. Pregnant women and individuals with sensitive skin or respiratory conditions should use monoterpene-rich essential oils with caution, ideally under the guidance of a healthcare professional.

It is important to store this particular group of oils in a dark, cool place away from heat and oxygen. This is because essential oils high in monoterpenes that have oxidized can cause skin irritation or sensitization. In addition, monoterpenes are not water-soluble and should not be used in the bath unless well diluted with a carrier oil or base. Limonene, present in all citrus oils, has been researched as a potential monoterpene risk as an oxidized irritant or allergen.

A few essential oils with a significant percentage of monoterpenes are:

- **Lemon (*Citrus limon*):** Rich in monoterpenes such as limonene, contributing to its bright, uplifting aroma and cleansing properties.
- **Pine (*Pinus sylvestris*):** Contains high levels of monoterpenes like α-pinene, which give it a fresh, woody scent and support respiratory health.
- **Peppermint (*Mentha piperita*):** Features monoterpenes such as menthol and

In essential oils, a functional group is a cluster of atoms that determines the oil's chemical behavior and therapeutic properties.

menthone, which provide a cooling effect and help with digestive and respiratory issues.

- **Eucalyptus (*Eucalyptus globulus*):** Known for its high content of monoterpenes like 1,8-cineole, offering respiratory and antiseptic benefits.
- **Tea Tree (*Melaleuca alternifolia*):** Contains monoterpenes, including 1,8-cineole and terpinene, contributing to its antimicrobial and immune-supporting properties.

Monoterpenols

Unlike monoterpene, which is a terpene, monoterpenol is an alcohol. Note the difference in the name ending in "ol." Monoterpenols have a ten-carbon backbone, like monoterpenes, but with a functional group. This specific group (-OH) is called hydroxyl. Monoterpenols usually fall in the middle note category, with many being floral and very aromatic. If you are wondering what a functional group is, it is a group of atoms in a molecule that usually goes through some chemical reaction and will distinguish its behavior. Interestingly, the molecule's therapeutic properties will be determined and changed based on the "-OH" location on the molecule.

All monoterpenols have this hydroxyl functional group (-OH), which is how someone can recognize it belongs to the monoterpenol chemical family. Alcohols are characterized by one or more hydroxyl groups (-OH), the oxygen atom bonded to the hydrogen structure. This is when size matters. If it's a 10-carbon, it's a monoterpenol. If it's a 15-carbon, it's a sesquiterpenol. Their shape resembles a monoterpene, except with a (-OH) attached. Depending on where the functional group attaches, different kinds of monoterpenols will be formed. Take geraniol, for instance. This primary alcohol is determined when the (-OH) attaches to a carbon atom linked to only one other carbon atom. Secondary alcohols are when two carbon atoms, such as menthol, are connected to the (-OH) carbon. Finally, tertiary alcohols are when three carbon atoms are attached to the (-OH) carbon. Terpinene-4-ol and linalool are examples of tertiary alcohols. These make good antimicrobials.

Essential oils such as clary sage, lavender, and rosewood contain the chemical component linalool. Research on linalool has found it effective

against bacteria (E. Joy Bowles). Linalool is also what makes clary sage and lavender quite sedative. Geraniol is present in palmarosa, geranium, and citronella and is responsible for its antifungal and anti-inflammatory properties. The component responsible for tea tree's antibacterial and antiviral qualities is Terpinen-4-ol, a member of the monoterpenol family. Menthol, which gives peppermint its cooling effect, which peppermint is notorious for, is also the same chemical molecule responsible for its analgesic and antispasmodic effect. Citronellol can be found in rose, lemon, eucalyptus, geranium, and other essential oils.

Common monoterpenols include:

- **Linalool:** Found in lavender, coriander, and basil, linalool is known for its calming and relaxing effects, as well as its potential antimicrobial properties.
- **Menthol:** In peppermint and spearmint oils, menthol provides a cooling sensation and is used for its soothing and analgesic effects on the skin and respiratory system.
- **Terpineol:** Found in pine oil and lilac, terpineol is valued for its pleasant fragrance and has antiseptic and calming properties.
- **Geraniol:** Common in rose and geranium oils, geraniol is recognized for its floral scent and is used for its antimicrobial and antioxidant properties.
- **Thymol:** Present in thyme oil, thymol has strong antimicrobial properties and is used for its ability to support respiratory health and act as a disinfectant.
- **Carvone:** Found in caraway and spearmint oils, carvone contributes to a fresh, minty aroma and has digestive and carminative properties.
- **Citronellol:** In citronella and rose oils, citronellol is known for its floral scent and is used

for its insect-repellent properties and its potential to relieve skin irritation.

Safety Considerations

Essential oils containing monoterpenols are non-toxic and mild on the skin and mucous membranes, except for peppermint, which can potentially be irritating. Do not use peppermint for children less than the age of five. In addition, infants should not be allowed to inhale peppermint. Essential oils rich in menthol should be avoided by anyone advised to avoid antimalarials, sulfonamides, chloramphenicol, streptomycin, or aspirin and in cases of cardiac fibrillation. Essential oils containing geraniol, such as geranium, citronella, rose, and thyme ct geraniol, should be diluted at 5%, as these oils can have mild skin-sensitizing effects. Monoterpenols are considered safe for long-term support for well-being and can help boost one's immune system. Essential oils containing monoterpenols usually have a three to five-year shelf life.

A few essential oils that contain monoterpenols include:

- **Lavender (*Lavandula angustifolia*):** Rich in linalool, known for its calming and relaxing properties.
- **Peppermint (*Mentha piperita*):** Contains menthol, which provides a cooling sensation and aids in respiratory and digestive support.
- **Rose (*Rosa damascena*):** High in geraniol, offering a floral fragrance and antimicrobial benefits.
- **Pine (*Pinus sylvestris*):** Includes terpineol, valued for its pleasant scent and antiseptic properties.

Oxides are essential oil compounds with notable antioxidant and antimicrobial properties.

- **Spearmint (*Mentha spicata*):** Contains menthol and carvone, known for their refreshing and digestive-supporting effects.
- **Geranium (*Pelargonium graveolens*):** Rich in geraniol, which contributes to its floral aroma therapeutic properties.
- **Rosemary (*Rosmarinus officinalis*):** Contains 1,8-cineole and camphor, contributing to its stimulating and respiratory-supporting effects.

These oils utilize their monoterpenol content to offer various therapeutic benefits, from relaxation and mood enhancement to respiratory support and antimicrobial action. Essential oils that contain a significant percentage of monoterpenols are typically very potent with anti-infectious properties such as antibacterial, antiviral, antifungal, and antimicrobial with airborne contaminants. Monoterpenols are mild enough for skincare and great as an antiseptic for healing wounds. Linalool offers a calming, sedative effect on the nervous system.

Oxides

Oxides are not a functional group but a group of compounds. Like ethers, oxides have an oxygen atom in the carbon chain, but it is usually combined in the shape of a ring. They are typically derived from alcohols with the -OH group modified as the hydrogen is removed and the oxygen bonds to another carbon to form a ring.

The most common oxide is 1,8-cineole in the chemical component family. Most essential oils contain 1,8-cineole, except for

Oxides in essential oils are known for their powerful respiratory benefits and antimicrobial properties.

German chamomile. Wikipedia states that 1,8-cineole is eucalyptol and is also known by various synonyms: 1,8-cineole, cajeputol, 1,8-epoxy-ethane, 1,8-oxido-p-menthane, eucalyptol, eucalyptol, 1,3,3-trimethyl-2-oxabicyclo [2,2,2] octane, cineol, cineole.

Oxides such as 1,8-cineole are a powerful respiratory aid and can have antiviral, antifungal, antibacterial, expectorant, and mucolytic effects. Essential oils containing oxides can act as expectorants as they stimulate mucus secretion and activate the cilia of the mucous membranes. When inhaled, these oils stimulate the brain and increase blood flow. Oils with 1,8-cineole can decrease shortness of breath and have an anti-inflammatory effect on bronchial asthma (see safety considerations).

Oxides in essential oils exhibit several therapeutic properties, including:

- **Expectorant:** Oxides, such as 1,8-cineole (eucalyptol), are known for their expectorant properties, helping to clear mucus from the respiratory tract and ease breathing in conditions like colds, flu, and bronchitis.
- **Anti-inflammatory:** Some oxides have anti-inflammatory effects, which can help reduce inflammation and alleviate symptoms of inflammatory conditions.
- **Antimicrobial:** Oxides possess antimicrobial properties, making them effective against bacteria, viruses, and fungi. This contributes to their use in disinfecting and preventing infections.
- **Analgesic:** Certain oxides can relieve pain, helping alleviate minor aches and pains through topical application or inhalation.
- **Bronchodilator:** Oxides like 1,8-cineole can act as bronchodilators, helping open the airways and improve respiratory function, which is particularly beneficial for asthma and other respiratory issues.

- **Stimulant:** Some oxides have stimulant properties, which can help enhance alertness, concentration, and overall mental clarity.
- **Decongestant:** Oxides effectively reduce nasal and sinus congestion and promote clearer breathing.
- **Immune Support:** Oxides can support the immune system, enhancing the body's ability to fight infections and maintain overall health.

Safety Considerations

People with asthma must use care with essential oils containing 1,8-cineole, as it could set off an attack. When choosing the oil, dilution, application, and circumstances should be considered. However, research shows that 1,8-cineole can help with respiratory disorders with its anti-inflammatory effects on passageways, possibly allowing a reduction in oral steroid dosage. Essential oils high in 1,8-cineole can oxidize, causing skin irritation and sensitization. It is also recommended not to use it on infants or children under ten. These oils may antidote homeopathic remedies. Do not use it with animals.

Essential oils that contain oxides include:

- **Eucalyptus (*Eucalyptus globulus*):** Rich in 1,8-cineole (eucalyptol), known for its decongestant and respiratory support properties.
- **Tea Tree (*Melaleuca alternifolia*):** Contains 1,8-cineole, contributing to its antimicrobial and antiseptic benefits.
- **Rosemary (*Rosmarinus officinalis*):** Includes 1,8-cineole, which helps with respiratory issues and mental clarity.
- **Cardamom (*Elettaria cardamomum*):** Contains 1,8-cineole, known for its digestive and respiratory support.

- **Nutmeg (*Myristica fragrans*):** Contains 1,8-cineole and other oxides, offering digestive and pain-relief benefits.
- **Laurel (*Laurus nobilis*):** Contains 1,8-cineole, which provides respiratory and digestive support.
- **Sage (*Salvia officinalis*):** Contains 1,8-cineole, which supports respiratory health and cognitive function.

Here are the major properties of 1,8-cineole:

- **Airborne Antimicrobial:** 1,8-cineole acts as an antimicrobial agent in the air, helping to reduce airborne pathogens.
- **Analgesic:** It has pain-relieving properties, making it helpful in alleviating discomfort.
- **Antibacterial:** Exhibits activity against bacterial infections, supporting its use in various health applications.
- **Antifungal:** Provides antifungal benefits, helping to combat fungal infections.
- **Anti-inflammatory:** Reduces inflammation, aiding in the relief of inflammatory conditions.
- **Antispasmodic:** Helps to relieve muscle spasms and cramping.
- **Antiviral:** Offers protection against viral infections, supporting overall immune health.
- **Expectorant:** Assists in clearing mucus from the respiratory tract, making it beneficial for respiratory health.
- **Increases Cerebral Blood Flow:** Enhances blood flow to the brain, improving cognitive function and mental clarity.
- **Mucolytic:** Breaks down mucus, facilitating easier expulsion from the respiratory system.
- **Reduces Tension Headaches:** Helps to alleviate tension headaches through its analgesic and anti-inflammatory properties.

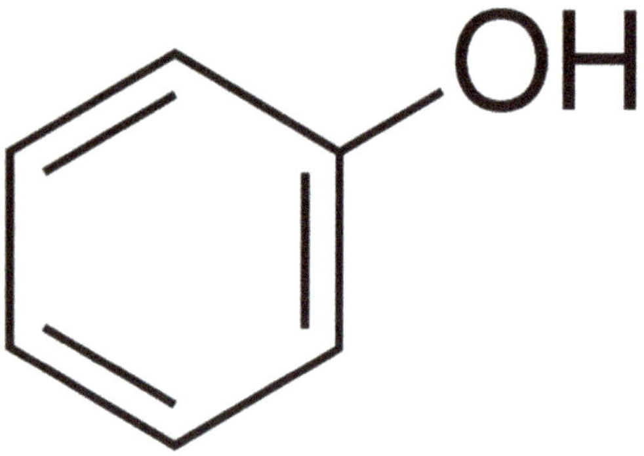

PHENOLS

- **Smooth Muscle Antispasmodic:** Relieves spasms in smooth muscles, such as those in the gastrointestinal tract.
- **Cough Suppressant:** Helps to reduce coughing by soothing the throat and respiratory passages.

Phenols

Phenols are known to be reactive as they release their hydrogen hydroxyl (-OH) and bind with other positively charged molecules, including protein, which can be harsh and cause skin damage. Essential oils containing phenols are aggressive and have a strong aroma. The benzene ring (6-carbon ring) distinguishes it from alcohols. The benzene ring is typically drawn with a circle in the middle to signify that all the bonds in the ring are equal. Like alcohols, many phenol names end in "-ol." Phenols tend to be base to middle notes, lasting longer on the skin. They are also responsible for the fragrance of an oil.

The shelf life of essential oils containing phenols tends to be three to five years.

Here are a few of the major properties of phenols:

- **Antimicrobial Activity:** Phenols are known for their strong antimicrobial properties, making them effective against bacteria, fungi, and viruses. This makes them useful in formulations for disinfectants and antibacterial treatments.

- **Antioxidant Effects:** Phenols possess potent antioxidant properties that help neutralize free radicals, reducing oxidative stress and potentially lowering the risk of chronic diseases.
- **Anti-inflammatory:** Many phenols have anti-inflammatory effects, which can help reduce inflammation and soothe conditions such as arthritis or skin irritations.
- **Analgesic:** Phenols can offer pain-relieving properties, relieving minor aches and pains when used in appropriate dilutions.
- **Stimulating:** Some phenols, such as those found in cinnamon, boost the immune system and can help enhance circulation and overall vitality.
- **Toxicity Concerns:** Due to their potent effects, phenols can be irritating or toxic in high concentrations. They should be used cautiously, particularly in individuals with sensitive skin, and always appropriately diluted.

- **Skin Sensitization:** Phenols can cause skin sensitization or irritation in some individuals, especially with prolonged or high-dose use. Patch testing is recommended before widespread application.

Safety Considerations

Essential oils high in phenols tend to irritate the skin and mucous membranes. It can be used with success when appropriately diluted. Using only at a 1% dilution with other skin-nourishing oils is recommended to balance your blend. Phenol-rich oils can be irritating to the skin in a bath. It should not be used on individuals with skin

Clove buds: potent with eugenol, offering strong antiseptic and analgesic benefits.

Phenols can stimulate the immune system when used for acute illnesses and infections.

issues or dermatitis. Essential oils high in eugenol, thymol, or carvacrol inhibit platelet aggregation. They should not be used by people with blood clotting disorders or individual anticoagulant drugs such as aspirin or warfarin before surgery.

Eugenol can be protective of the liver, but only in low concentrations. In high concentrations, it is hepatotoxic and can cause damage to tissues. Phenols can stimulate the immune system when used in acute illnesses and infections. Phenols are strengthening, analgesic, and anti-inflammatory. Thymol found in thyme is beneficial for acute respiratory infections. Clove bud essential oil is 67% phenols and is excellent for combating infections, but people on blood thinners should avoid it due to its high eugenol content. They contain high levels of oxygenating molecules and have antioxidant properties.

Oils that contain Phenols:

- **Clove (*Syzygium aromaticum*)**: Rich in eugenol, known for its powerful antiseptic and analgesic properties.
- **Thyme (*Thymus vulgaris*)**: The thymol chemotype is high in thymol and offers strong antibacterial and antifungal properties.
- **Oregano (*Origanum vulgare*)**: Contains carvacrol and thymol, both effective against bacteria and fungi.
- **Basil (*Ocimum basilicum*)**: Certain chemotypes contain eugenol, contributing to its antimicrobial and anti-inflammatory effects.
- **Savory (*Satureja montana*)**: High in carvacrol and thymol, similar to oregano, known for strong antimicrobial action.

- **Cinnamon (*Cinnamomum zeylanicum* or *Cinnamomum cassia*)**: Contains cinnamaldehyde and eugenol, providing strong antiseptic and antimicrobial properties.
- **Anise (*Pimpinella anisum*)**: Contains anethole, a phenolic compound with antimicrobial and antifungal properties.

Sesquiterpenes

Sesquiterpenes comprise three isoprene units with five carbon atoms each, totaling fifteen total carbon atoms. Each one has varying hydrogen atoms depending on how many double bonds there are. Sesquiterpenes and monoterpenes are made up of only carbon and hydrogen. The name of sesquiterpene molecules ends in "ene." Essential oils high in sesquiterpene tend to be more emotionally grounding and calming. These oils are heavier, typically middle to base notes with a longer-lasting aroma. This is because sesquiterpenes decrease their volatility (evaporation rate).

Sesquiterpenes are known to work as liver and gland stimulants because of the chemical components caryophyllene and valencene. Studies have shown that sesquiterpenes can pass the blood-brain barrier, enter brain tissue, and increase oxygen around the pineal and pituitary glands. Sesquiterpenes such as chamazulene and farnesol are highly anti-inflammatory and antibacterial. At the same time, farnesol is known to be antiviral in action. Chamazulene is found in German chamomile, tansy, and yarrow.

The sesquiterpene chemical family is found in some essential oils, with over 50% of the oil being sesquiterpenes. Caryophyllene gives a robust, woody scent to many essential oils such as clove,

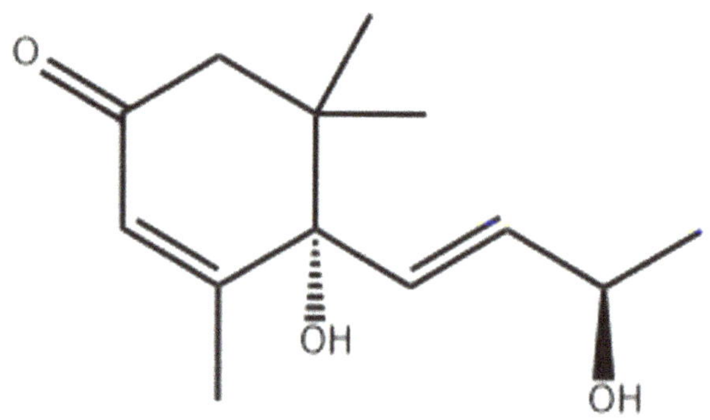

SESQUITERPENES

lavender, thyme, black pepper, melissa, cinnamon, and ylang ylang.

Some sesquiterpenes are unique to a single oil. Sesquiterpenes also increase the half-life of smaller light molecules, both aromatically and therapeutically. These are called "fixing oils" and are frequently used by perfumers to help make a fragrance last longer. Essential oils such as myrrh and sandalwood both contain over 90% sesquiterpenes.

Essential oils high in sesquiterpenes have a long shelf life, typically six to eight years or longer.

> *Caryophyllene imparts a rich, woody aroma to essential oils like clove, lavender, thyme, black pepper, melissa, cinnamon, and ylang ylang.*

Sesquiterpenes in essential oils exhibit several therapeutic properties, including:

- **Anti-inflammatory:** Sesquiterpenes have potent anti-inflammatory effects, helping to reduce inflammation and alleviate symptoms associated with inflammatory conditions such as arthritis and dermatitis.
- **Analgesic:** They provide pain-relieving properties, aiding in reducing minor aches, pains, and muscle soreness.
- **Antimicrobial:** Sesquiterpenes possess antimicrobial properties, making them effective in inhibiting the growth of bacteria, fungi, and viruses, which helps prevent infections.
- **Antispasmodic:** They can help relieve muscle spasms and cramps, promoting relaxation and reducing discomfort.
- **Sedative:** Certain sesquiterpenes have calming and soothing effects, which can help reduce anxiety, promote relaxation, and improve sleep quality.

- **Antioxidant:** Sesquiterpenes exhibit antioxidant properties, neutralizing free radicals and reducing oxidative stress, which supports overall health and may help prevent chronic diseases.
- **Immune Support:** They can enhance immune function, helping the body to better defend against infections and illnesses.
- **Skin Benefits:** Sesquiterpenes support skin health by promoting cell regeneration, reducing inflammation, and soothing irritated skin, making them beneficial for treating various skin conditions.
- **Decongestant:** Some sesquiterpenes have decongestant properties, helping to relieve nasal and chest congestion and promoting clearer breathing.

Safety Considerations

Properly dilute before topical application to prevent skin irritation or sensitization. Despite their generally gentle nature, some individuals may still experience adverse reactions, so a patch test is recommended before extensive use. Sesquiterpene-rich oils, such as cedarwood, sandalwood, and ginger, are often used for their calming and anti-inflammatory properties. Still, they should be used with caution during pregnancy and breastfeeding, as their effects in these conditions are not fully understood. Additionally, while these oils are typically considered safe for most adults, those with sensitive skin or respiratory conditions should consult a healthcare professional before use. Inhalation of sesquiterpene-rich oils should be done in well-ventilated areas to avoid potential respiratory irritation. Sesquiterpenes are

generally safe and not skin irritants unless oils have oxidized.

Oils that contain Sesquiterpenes:

- **Cedarwood (*Cedrus atlantica*)**: Contains cedrol and other sesquiterpenes, which provide grounding and calming effects.
- **Sandalwood (*Santalum album*)**: Rich in α-santalol and β-santalol, known for their relaxing and anti-inflammatory properties.
- **Ginger (*Zingiber officinale*)**: Contains zingiberene, which offers anti-inflammatory and digestive benefits.
- **Patchouli (*Pogostemon cablin*)**: High in patchoulol, contributing to its grounding and skin-healing properties.
- **Myrrh (*Commiphora myrrha*)**: Contains furanodiene and curzerene, known for their anti-inflammatory and antimicrobial effects.
- **Chamomile (*Matricaria recutita* or *Chamaemelum nobile*)**: Contains bisabolol, which has anti-inflammatory and soothing properties.
- **Black Pepper (*Piper nigrum)***: Contains β-caryophyllene, offering anti-inflammatory and analgesic benefits.
- **Vetiver (*Vetiveria zizanioides*)**: Rich in vetiverol and khusimol, known for their grounding and calming effects.
- **Frankincense (*Boswellia carterii*)**: Contains α-pinene and other sesquiterpenes, known for their anti-inflammatory and immune-boosting properties.
- **Ylang Ylang (*Cananga odorata*)**: Contains caryophyllene, contributing to its sedative and calming effects.

Sesquiterpenols

Like sesquiterpenes, sequiterpenols have a 15-carbon backbone with added hydroxyl functional group (-OH). Essential oils high in sequiterpenols are considered grounding and tend to be heavier, consisting primarily of base notes.

One distinctive trait of individual plants that contain sesquiterpenols is that they create sesquiterpenols unique to that plant. For instance, patchoulol found in patchouli is rarely found in other oils, as is santalol, which is only in sandalwood. Sandalwood is 85% sesquiterpenols. Components in the sesquiterpenol family end in "-ol."

The sesquiterpenol chemical family is found in certain essential oils like sandalwood and cedarwood, where it plays a key role in stabilizing and prolonging the aroma of lighter, more volatile molecules. Known as "fixing oils," sesquiterpenols help extend the lifespan of fragrances, making them popular in perfumery. Some sesquiterpenols are unique to specific oils and offer distinct therapeutic benefits. Oils high in sesquiterpenols, such as sandalwood, typically have a shelf life of six to eight years due to their chemical stability.

Sesquiterpenols in essential oils exhibit several therapeutic properties, including:

- **Anti-inflammatory:** Sesquiterpenols have potent anti-inflammatory effects, helping to reduce inflammation and alleviate symptoms associated with inflammatory conditions.
- **Antimicrobial:** They possess antimicrobial properties, making them effective against various bacteria, fungi, and viruses, which help prevent infections.

- **Analgesic:** Sesquiterpenols provide pain-relieving properties, aiding in the reduction of minor aches, pains, and muscle soreness.
- **Sedative:** Certain sesquiterpenols have calming and soothing effects, which can help reduce anxiety, promote relaxation, and improve sleep quality.
- **Skin Regeneration:** They support skin health by promoting cell regeneration and repair, which is beneficial for treating wounds, scars, and various skin conditions.
- **Immune Support:** Sesquiterpenols can enhance immune function, helping the body to better defend against infections and illnesses.
- **Antioxidant:** They exhibit antioxidant properties, neutralizing free radicals and reducing oxidative stress, contributing to overall health and potentially preventing chronic diseases.
- **Decongestant:** Some sesquiterpenols have decongestant properties, helping to relieve nasal and chest congestion and promoting clearer breathing.
- **Balancing:** Sesquiterpenols are known for their balancing effects, which can help regulate various bodily functions and maintain homeostasis.

> *Sesquiterpenols, found in essential oils, are celebrated for their calming, anti-inflammatory, and regenerative effects on both body and mind.*

Safety Considerations

Sesquiterpenols, a subgroup of sesquiterpenes found in various essential oils, are generally considered safe due to their low toxicity and non-irritating nature. Essential oils containing sesquiterpenols are usually suitable for topical applications, as they are less likely to cause skin irritation or sensitization. However, diluting these oils with a carrier oil before use is crucial to minimize any potential irritation. Conducting a patch test before using a new blend can help ensure no adverse reaction.

While sesquiterpenols are typically safe, it is advisable to consult with a healthcare provider before using them during pregnancy or breastfeeding. Additionally, caution should be exercised when using essential oils with sesquiterpenols on children, ensuring proper dilution and avoiding oils not recommended for pediatric use.

However, β-eudesmol found in blue cypress is cautioned against anticoagulant drugs, peptic ulcers, or any bleeding disorders. Blue cypress should also be avoided during pregnancy.

Oils that contain Sesquiterpenols:

- **Sandalwood (*Santalum album*):** Rich in the sesquiterpenol santalol, known for its calming and grounding properties.
- **Cedarwood (*Cedrus atlantica*):** Contains cedrol, a sesquiterpenol with comforting and antiseptic qualities.
- **Patchouli (*Pogostemon cablin*):** Features patchoulol, a sesquiterpenol valued for its skin-soothing and regenerative effects.
- **Ginger (*Zingiber officinale*):** Includes zingerone, a sesquiterpenol recognized for its warming and anti-inflammatory properties.
- **Frankincense (*Boswellia serrata*):** Contains boswellic acids and other sesquiterpenols that offer a range of therapeutic benefits.
- **Myrrh (*Commiphora myrrha*):** Rich in sesquiterpenols such as curzerene, contributing to its anti-inflammatory and soothing properties.
- **Helichrysum (*Helichrysum italicum*):** Known for its high content of sesquiterpenols like italidiones, which support skin regeneration and healing.
- **Vetiver (*Vetiveria zizanoides*):** Contains vetiverol, a sesquiterpenol that is grounding and supports emotional well-being.

CHEMICAL COMPOSITION

Study the charts below to select oils based on their chemical composition for specific therapeutic benefits. Keep this list handy for creating your own essential oil datasheets.

MONOTERPENES	
Camphene	antioxidant, mucolytic
Carene (δ-3)	anti-inflammatory, inhibits acetylcholinesterase, mucolytic, prevents bone loss
Cymene (para)	analgesic, antibacterial, antiviral
Limonene (d)	activates white blood cells, analgesic, anti-inflammatory, antibacterial, antiobesity, antioxidant, antitumoral, antiulcerogenic, immunostimulant, skin penetration enhancer
Myrcene (β)	analgesic, anti-inflammatory, sedative
Phellandrene (α)	antibacterial, antispasmodic
Phellandrene (β)	antifungal (candida)
Pinene (α)	anti-inflammatory, antibacterial, antifungal (candida), antispasmodic, antiviral, prevents bone loss
Pinene (β)	anti-inflammatory, antioxidant, antispasmodic, prevents bone loss
Terpinene (γ)	antibacterial, antioxidant, antispasmodic, antiviral, kills head lice
Terpinolene	antibacterial, antifungal (candida), antioxidant, antispasmodic, antiviral, sedative

CHEMICAL COMPOSITION

SESQUITERPENES	
Caryophyllene (α)	anti-inflammatory, antitumoral
Caryophyllene (β)	analgesic, anti-inflammatory, antispasmodic, antitumoral, antiviral, immunostimulant
Cedrene	skin penetration enhancer
Curcumene (α)	anti-inflammatory, antitumoral, antiulcerogenic, lowers triglycerides
Sesquiphellandrene (β)	antitumoral, antiviral
Zingiberene	antiulcerogenic

MONOTERPENOLS	
Borneol	anti-inflammatory, antibacterial, anticoagulant, antifungal, antioxidant, antispasmodic, antiviral, cicatrisant, cns depressant, inhibits acetylcholinesterase, neuroprotective, prevents bone loss, sedative
Citronellol	anti-inflammatory, antianxiety, anticonvulsant, antifungal, antifungal (candida), antioxidant, cns depressant, hypotensive, kills dust mites
Geraniol	airborne antimicrobial, anti-inflammatory, antibacterial, antifungal, antifungal (candida), antioxidant, antitumoral, cooling, kills dust mites
Linalool	airborne antimicrobial, analgesic, anti-inflammatory, anti-anxiety, antibacterial, anticonvulsant, antifungal (candida), antioxidant, antispasmodic, antitumoral, antiviral, hypotensive, immunostimulant, sedative

CHEMICAL COMPOSITION

MONOTERPENOLS	
Menthol	analgesic, anti-inflammatory, antibacterial, anticoagulant, antioxidant, antipruritic, antispasmodic, antitumoral, antitussive, choleretic, cns stimulant, cooling, dopaminergic, hypotensive
Terpinen-4-ol	activates white blood cells, anti-inflammatory, antibacterial, anticonvulsant, antifungal, antifungal (candida), antispasmodic, antitumoral, antiulcerogenic, antiviral, cns depressant, hypotensive, immunostimulant, kills head lice, treats scabies
Terpineol (α)	airborne antimicrobial, analgesic, anti-inflammatory, antibacterial, antifungal, antioxidant, antitumoral, antiviral, cicatrisant, hypotensive, kills head lice, sedative

ETHERS	
Anethole (E)	analgesic, anti-inflammatory, anticoagulant, antifungal, antithrombotic, antiulcerogenic, antiviral, inhibits acetylcholinesterase, sedative

ALDEHYDES	
Citral (geranial and neral)	airborne antimicrobial, analgesic, anti-inflammatory, antibacterial, anticonvulsant, antifungal, antifungal (candida), antispasmodic, antitumoral, antiviral, sedative
Citronellal	airborne antimicrobial, analgesic, antibacterial, antioxidant, antitumoral, cns depressant, sedative

CHEMICAL COMPOSITION

SESQUITERPENOLS	
Bisabolol (α)	anti-inflammatory, antifungal, antimicrobial, antioxidant, antiparasitic, antispasmodic, antitumoral, antiulcerogenic, cicatrisant, skin penetration enhancer
Cedrol	antibacterial, antifungal, antiobesity, antitumoral, hypotensive, paf receptor binding antagonist, sedative, skin penetration enhancer
Eudesmol (β)	aids digestion, anti-inflammatory, antifungal, antitumoral, cns depressant, hypotensive, liver protective
Nerolidol	anti-inflammatory, antibacterial, antifungal, antispasmodic, antitumoral, antiulcerogenic
Patchoulol	antispasmodic, sedative
Santalols	analgesic, antibacterial, antifungal (candida), antimicrobial, antitumoral, improves mood, neuroleptic, sedative

ESTERS	
Benzyl benzoate	antifungal (candida), antispasmodic, antitumoral, hypotensive, kills dust mites, kills head lice, treats scabies
Bornyl acetate	analgesic, anti-inflammatory, cns depressant, prevents bone loss, sedative
Geranyl acetate	antibacterial, antifungal (candida), choleretic
Linalyl acetate	analgesic, anti-inflammatory, antibacterial, antitumoral, antiulcerogenic, immunostimulant, sedative
Methyl cinnamate	antibacterial, antifungal, antioxidant

CHEMICAL COMPOSITION

KETONES	
Camphor	analgesic, antitussive, cns stimulant, mucolytic, prevents bone loss
Carvone	activates white blood cells, analgesic, anti-inflammatory, anti-anxiety, antibacterial, anticonvulsant, antifungal, antispasmodic, antitumoral, cns depressant, immunostimulant, liver protective, sedative
Menthone	analgesic, anti-inflammatory, antibacterial, anticoagulant, antioxidant, cns stimulant
Turmerones	anti-inflammatory, anticoagulant, antifungal, antitumoral

PHENOLS	
Carvacrol	airborne antimicrobial, analgesic, anti-inflammatory, anti-anxiety, antibacterial, antifungal, antifungal (candida), antioxidant, antispasmodic, antitumoral, hypotensive, inhibits acetylcholinesterase, liver protective
Eugenol	analgesic, anti-inflammatory, antibacterial, anticoagulant, anticonvulsant, antifungal (candida), antihistaminic, antioxidant, antispasmodic, antithrombotic, antitumoral, antiulcerogenic, antiviral, hypotensive, liver protective, neuroprotective, vasodilator
Thymol	analgesic, anti-inflammatory, antibacterial, antifungal, antifungal (candida), antioxidant, antispasmodic, antitumoral, antiviral, hypotensive, liver protective

CHEMICAL COMPOSITION

OXIDES	
Bisabolol oxides (α)	anti-inflammatory, antimicrobial, antispasmodic
Cineole	airborne antimicrobial, analgesic, anti-inflammatory, antibacterial, antioxidant, antispasmodic, antiviral, dopaminergic, hypotensive, increases cerebral blood flow, inhibits acetylcholinesterase, mucolytic

ACIDS	
Benzoic Acid	antibacterial, antifungal, anti-inflammatory, expectorant, antiseptic, wound healing, preservative
Rosmarinic Acid	anti-inflammatory, antioxidant, antibacterial, antiviral, anti-allergic, neuroprotective, anti-carcinogenic, liver protective
Cinnamic Acid	antibacterial, antifungal, antiviral, antioxidant, anti-inflammatory, antitumoral, astringent, immune-stimulating
Salicylic Acid	analgesic, anti-inflammatory, antibacterial, antifungal, keratolytic, anti-acne, exfoliant, antipyretic

OTHER CHEMICAL	
Chamazulene	anti-inflammatory, antibacterial, antioxidant, antispasmodic, liver protective

Blending essential oils based on chemistry involves combining oils by their chemical components to create harmonious, effective, and safe mixtures tailored to specific therapeutic and aromatic goals.

"

Essential oils are complex mixtures of volatile compounds, primarily terpenes, esters, alde-hydes, ketones, alcohols, phenols, and oxides.

CHAPTER FIVE

The Importance of Aromatherapy

Aromatherapy, the practice of using essential oils for therapeutic benefits, has roots stretching back thousands of years. Despite its ancient origins, the modern understanding of aromatherapy hinges on the principles of chemistry. Understanding the chemical composition of essential oils is crucial for several reasons: it informs their therapeutic properties, safety, and efficacy and guides their appropriate use.

Essential oils are complex mixtures of volatile compounds, primarily terpenes, esters, aldehydes, ketones, alcohols, phenols, and oxides. Each of these compounds contributes to the oil's overall therapeutic profile. For example:

Terpenes: These hydrocarbons are essential oils' largest class of compounds. They are known for their anti-inflammatory and antiseptic properties. Limonene, a common terpene found in citrus oils, has been shown to possess mood-enhancing and immune-stimulating effects.

Esters: Characterized by their pleasant, often fruity aroma, esters are known for their calming and antispasmodic properties. Linalyl acetate, found in lavender oil, is a prominent ester contributing to the oil's soothing effects.

Phenols: These compounds are known for their potent antimicrobial and antioxidant properties. Thymol, found in thyme oil, is a well-known phenol with significant antiseptic qualities.

By understanding the chemical makeup of essential oils, practitioners can predict and harness their therapeutic effects more accurately.

Safety and Efficacy: The Chemical Perspective

Safety is paramount in aromatherapy, and chemistry plays a critical role in ensuring the safe use of essential oils. Some essential oils contain compounds that can cause skin irritation, sensitization, or phototoxic reactions. For instance, furocoumarins in some citrus oils like bergamot can cause severe skin reactions when exposed to sunlight.

Moreover, certain compounds can be toxic in high concentrations. For example, pulegone in pennyroyal oil and safrole in sassafras oil are known hepatotoxins and must be used cautiously. Understanding these chemical properties allows practitioners to recommend safe dilution ratios and application methods.

Chemotypes and Their Significance

A chemotype is a chemically distinct entity within a plant species, distinguished by the composition of its essential oil. Chemotypes are crucial in aromatherapy because they highlight how the same plant can produce oils with different chemical profiles based on environmental factors like soil, climate, and altitude.

For example, *Thymus vulgaris* (thyme) has several chemotypes, such as thymol, linalool, and geraniol chemotypes. Each chemotype has a unique

chemical composition and different therapeutic properties. Understanding chemotypes ensures that the correct type of oil is selected for specific therapeutic outcomes.

Synergy and Blending: The Chemistry Behind Effective Formulations

One of the fundamental practices in aromatherapy is blending essential oils to create synergistic effects. Synergy occurs when the combined effect of individual essential oils is greater than the sum of their effects when used separately. Chemistry helps elucidate these interactions and guide effective blending practices.

For example, blending oils with complementary chemical constituents can enhance their therapeutic effect. A blend of lavender (rich in esters like linalyl acetate) and peppermint (high in menthol monoterpene alcohol) can offer enhanced analgesic and anti-inflammatory properties.

Analytical Techniques in Essential Oil Chemistry

The study of essential oil chemistry relies on various analytical techniques to identify and quantify their components. Gas chromatography (GC) and mass spectrometry (MS) are the most commonly used methods. These techniques allow for precisely identifying individual compounds within an essential oil, providing a detailed chemical profile.

Understanding these analytical techniques is essential for practitioners to ensure the purity

Enfleurage captures delicate floral aromas' true essence.

and quality of essential oils. Adulteration and contamination are significant concerns in the industry, and chemistry-based analysis helps verify essential oils' authenticity and therapeutic value.

The importance of chemistry in aromatherapy cannot be overstated. It provides the foundation for understanding essential oils' therapeutic properties, safety, and efficacy. A solid grasp of essential oil chemistry enables practitioners to make informed decisions, ensuring that aromatherapy is effective and safe. As the field continues to evolve, ongoing research into the chemical composition of essential oils will undoubtedly lead to new insights and applications, further solidifying the bond between chemistry and aromatherapy.

Significance in Aromatherapy

The detailed study of molecular structures and functional groups allows aromatherapists to select and combine essential oils precisely. By understanding the specific actions of different chemical constituents, practitioners can tailor their approaches to meet individual therapeutic needs, ensuring both efficacy and safety. This knowledge also facilitates the creation of synergistic blends, where the combined effects of multiple oils produce enhanced therapeutic benefits.

In summary, exploring molecular structures and functional groups provides invaluable insights into the chemical nature of essential oils. This understanding is essential for leveraging their full potential in therapeutic applications, ensuring that the oils are used effectively and safely. As we delve deeper into this book, we will break down the key molecular structures and functional groups, examining their roles and implications in greater detail.

By understanding the specific actions of different chemical constituents, practitioners can tailor their approaches to meet individual therapeutic needs, ensuring both efficacy and safety.

"

Chemical compounds and constituents in essential oils define their unique properties, therapeutic effects, and aromatic profiles, making them essential for crafting precise and effective blends.

CHAPTER SIX

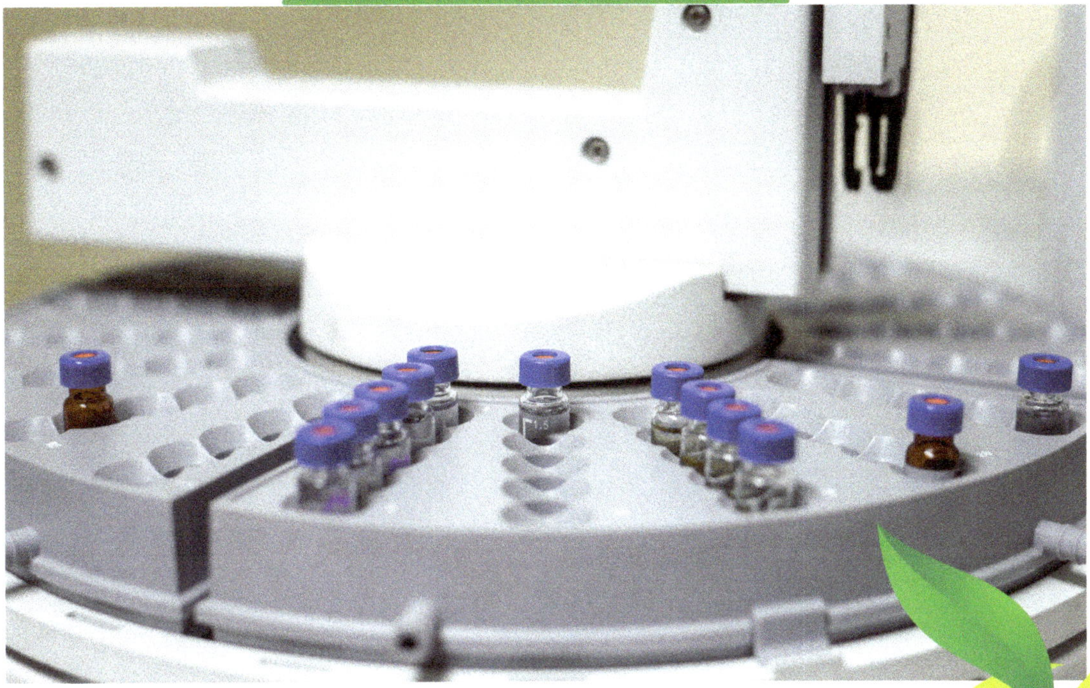

The Importance of Chemistry in Understanding Essential Oils

Essential oils contain a complex amalgamation of natural compounds with a rich and diverse chemistry underpinning their therapeutic and aromatic properties. These volatile substances are primarily composed of various organic compounds, including terpenes, alcohols, esters, and ketones, each contributing to their distinct scent and potential health benefits. Terpenes, such as limonene and linalool, often dominate essential oils and are responsible for their characteristic fragrances. Additionally, these compounds exhibit a wide range of biological activities, from anti-inflammatory and antiseptic effects to mood enhancement and stress reduction. The intricate chemistry of essential oils makes them valuable in aromatherapy and as natural remedies for numerous ailments, stimulating the senses and promoting overall well-being.

The average essential oil contains around 100 to 300 components, along with thousands of other trace compounds that have yet to be identified by scientists. While attempts have been made to single out and synthetically duplicate identifiable components, nature far exceeds laboratory efforts and remains a mystery in its unique makeup.

The specific chemistry of an essential oil is influenced by such a diverse range of factors that it is impossible to recreate its components in a

Identifying an essential oil's chemical composition ensures its authenticity, quality, and purity.

laboratory synthetically. The chemical composition of an essential oil can be affected by its climatic conditions, pollutant exposure, the extraction process, and other factors that can influence its aroma and therapeutic powers.

Chemistry plays a pivotal role in understanding essential oils and their myriad benefits. Essential oils are complex mixtures of volatile compounds produced by plants, each with distinct chemical structures and properties. By studying the chemistry of these oils, we can gain insight into their therapeutic effects, safety profiles, and applications in aromatherapy and beyond.

Identifying Active Components

Each essential oil comprises numerous chemical constituents, including terpenes, alcohols, esters, aldehydes, ketones, phenols, and oxides. The specific combination of these components determines the oil's unique aroma, therapeutic properties, and potential applications. For example, the calming effects of lavender essential oil are primarily due to its high concentration of linalool and linalyl acetate. Understanding these active components helps practitioners select the appropriate oils for specific therapeutic goals.

Predicting Therapeutic Effects

The therapeutic effects of essential oils are determined mainly by their chemical composition. For instance, oils rich in monoterpenes, such as limonene in citrus oils, are known for their antiviral and uplifting properties. Sesquiterpenes, found in oils like sandalwood and cedarwood, have anti-inflammatory and calming effects. By analyzing the chemical makeup of an essential oil, we can predict its potential benefits and tailor treatments to meet individual needs.

Ensuring Safety and Efficacy

Understanding the chemistry of essential oils is crucial for ensuring their safe and effective use. Some chemical constituents can be irritating or toxic if misused. For instance, phenols such as thymol and carvacrol are potent antimicrobials but can cause skin irritation and liver toxicity if used in high concentrations or over extended periods. Knowledge of these properties enables practitioners to use essential oils safely, selecting appropriate dilutions and application methods to avoid adverse reactions.

Enhancing Blending Techniques

Chemistry also informs the art and science of blending essential oils. By understanding the chemical profiles of different oils, practitioners can create synergistic blends that enhance desired effects. For example, blending oils high in monoterpenes with those rich in sesquiterpenes can provide a balanced combination of uplifting and grounding properties. This knowledge allows for creating customized blends tailored to individual therapeutic needs.

Improving Product Quality

The quality and purity of essential oils are directly related to their chemical composition. Adulteration, contamination, or poor extraction methods can alter the chemical profile of an essential oil, reducing its efficacy and safety. Analytical techniques such as gas chromatography-mass spectrometry (GC-MS) are used to verify the chemical composition of essential oils, ensuring their purity and potency. Understanding these

analytical methods is necessary for practitioners to assess and ensure the quality of their oils.

Ongoing research into the chemistry of essential oils continues to uncover new therapeutic applications and mechanisms of action.

Advancing Research and Development

Ongoing research into the chemistry of essential oils continues to uncover new therapeutic applications and mechanisms of action. Advances in analytical techniques and molecular biology provide deeper insights into how essential oil components interact with biological systems. This research expands our understanding of essential oils' potential benefits and paves the way for new medical and wellness applications.

In summary, a solid understanding of chemistry is fundamental to the effective and safe use of essential oils. It allows practitioners to identify active components, predict therapeutic effects, ensure safety, enhance blending techniques, improve product quality, and advance research. By appreciating the chemical complexity of essential oils, practitioners can harness their full potential for promoting health and well-being.

Chemical Properties and Their Influence on Aroma

The distinctive aromas of essential oils are directly influenced by their chemical composition, specifically by the types and proportions of volatile compounds they contain. Understanding the chemical properties responsible for these aromas provides insight into how essential oils are categorized, blended, and used in aromatherapy and perfumery.

Volatility and Molecular Weight

The volatility of an essential oil—its ability to evaporate at room temperature—is primarily determined by the molecular weight of its components. Lighter molecules, such as monoterpenes and their derivatives, are highly volatile and evaporate quickly, releasing their aromas into the air. In contrast, heavier molecules, like sesquiterpenes and diterpenes, evaporate more slowly and contribute to the base notes of an oil's aroma profile. The balance between volatile and less volatile components shapes an essential oil's overall aroma intensity and longevity.

Functional Groups and Aromatic Profiles

The functional groups in essential oil constituents are pivotal in defining their aromatic profiles. For example, alcohols like linalool and geraniol contribute to floral and sweet aromas, while aldehydes like citral impart bright, citrusy notes. Ketones like camphor and menthone often contribute to cooling or minty aromas, while phenols like thymol and carvacrol can have strong, medicinal scents. Esters such as linalyl acetate lend fruity and floral nuances to essential oils. The specific combination and concentration of these functional groups determine the overall aroma character of an oil.

The aromatic characteristics of essential oils are intricately linked to their chemical properties.

Chirality and Aromatic Complexity

Chirality, or asymmetry in molecular structure, adds another layer of complexity to essential oil aromas. Many essential oil components exist in chiral forms, where molecules are mirror images of each other (enantiomers). These enantiomers can have subtly different aromas and therapeutic properties. For instance, the two enantiomers of limonene may smell similar but can interact differently with olfactory receptors or have varied effects in the body. Chirality enhances the richness and depth of an essential oil's aromatic profile, contributing to its complexity.

Chemical Interactions and Synergy

The interactions between different chemical constituents within an essential oil can create aromatic synergy. Synergy occurs when the combined effects of multiple compounds produce a more powerful or pleasing aroma than each compound alone. This phenomenon is commonly observed in well-balanced essential oil blends, where complementary aromas harmonize to create a unique olfactory experience. Understanding these chemical interactions allows aromatherapists and perfumers to craft blends that achieve desired aromatic effects for relaxation, stimulation, or emotional balance.

Stability and Aging

The chemical stability of essential oil constituents influences how their aromas evolve. Some compounds are more prone to oxidation or degradation, which can alter their aromatic profiles. For example, citrus oils high in monoterpenes like limonene may become less vibrant and more resinous with age. In contrast, sesquiterpene-rich oils such as sandalwood can develop deeper, decadent aromas over time. Proper storage and handling practices, such as using dark glass bottles and storing oils away from heat and light, help maintain the integrity of essential oils and preserve their aromatic qualities.

In summary, the aromatic characteristics of essential oils are intricately linked to their chemical properties. By understanding these properties—such as volatility, functional groups, chirality, chemical interactions, and stability—practitioners can effectively harness the aromatherapeutic potential of essential oils, creating blends and formulations that evoke specific moods, support well-being, and enhance sensory experiences.

Analytical Techniques in Essential Oil Chemistry

Understanding the complex chemical composition of essential oils, which dictates their aroma, therapeutic benefits, and overall quality, relies on advanced analytical techniques. This section explores key methods and instruments critical in studying essential oil chemistry.

Gas Chromatography-Mass Spectrometry (GC-MS)

Gas Chromatography-Mass Spectrometry (GC-MS) is the cornerstone of essential oil analysis. It separates the volatile components of oils based on their molecular weight and polarity, identifying and quantifying each compound by comparing its mass spectrum to a database of known substances. GC-MS provides detailed insights into the presence and proportions of terpenes, alcohols, esters, and other volatile compounds, crucial for understanding an oil's composition.

Fourier Transform Infrared Spectroscopy (FTIR)

Fourier Transform Infrared Spectroscopy (FTIR) analyzes how molecules in essential oils absorb infrared radiation, revealing information about their functional groups—such as alcohols, aldehydes, ketones, and esters. FTIR confirms the presence of specific constituents by detecting characteristic chemical bonds, aiding in the comprehensive characterization of essential oils.

High-Performance Liquid Chromatography (HPLC)

High-Performance Liquid Chromatography (HPLC) complements GC-MS by separating and quantifying individual compounds in essential oils based on their chemical properties and affinity for a stationary phase. Particularly effective for non-volatile or semi-volatile substances like phenols and

Analytical techniques like GC-MS, FTIR, and HPLC are vital for decoding the complex composition of essential oils.

acids, HPLC offers additional insights into the diverse chemical profile of oils that GC-MS alone may not capture fully.

Nuclear Magnetic Resonance (NMR) Spectroscopy

Nuclear Magnetic Resonance (NMR) Spectroscopy elucidates the molecular structure and dynamics of essential oil components by detecting the magnetic properties of atomic nuclei. This technique identifies complex molecules, determines their spatial arrangements, and uncovers interactions within molecular environments. NMR is invaluable for studying compound chirality and understanding their biological roles within essential oils.

Mass Spectrometry (MS)

Mass Spectrometry (MS) measures the mass-to-charge ratio of ionized molecules in essential oils, providing precise data on their molecular weight and fragmentation patterns. When coupled with separation techniques like GC or liquid chromatography (LC), MS enhances the identification and quantification of compounds, especially those present in trace amounts. Advanced MS variants, such as tandem mass spectrometry (MS/MS), offer enhanced sensitivity and specificity for complex oil matrices.

Additional Techniques and Applications

Techniques such as polarimetry, refractometry, and colorimetry contribute to essential oil analysis by measuring optical rotation, refractive index, and color changes due to oxidation. Emerging technologies like chemometrics and molecular modeling further advance predictive modeling of aroma profiles and biological activities based on chemical data.

Analytical techniques in essential oil chemistry are indispensable for uncovering oils' chemical complexity, purity, and functional properties. These methods empower researchers, producers, and regulatory bodies to ensure essential oils' consistency, safety, and efficacy across various applications—from aromatherapy and cosmetics to pharmaceuticals.

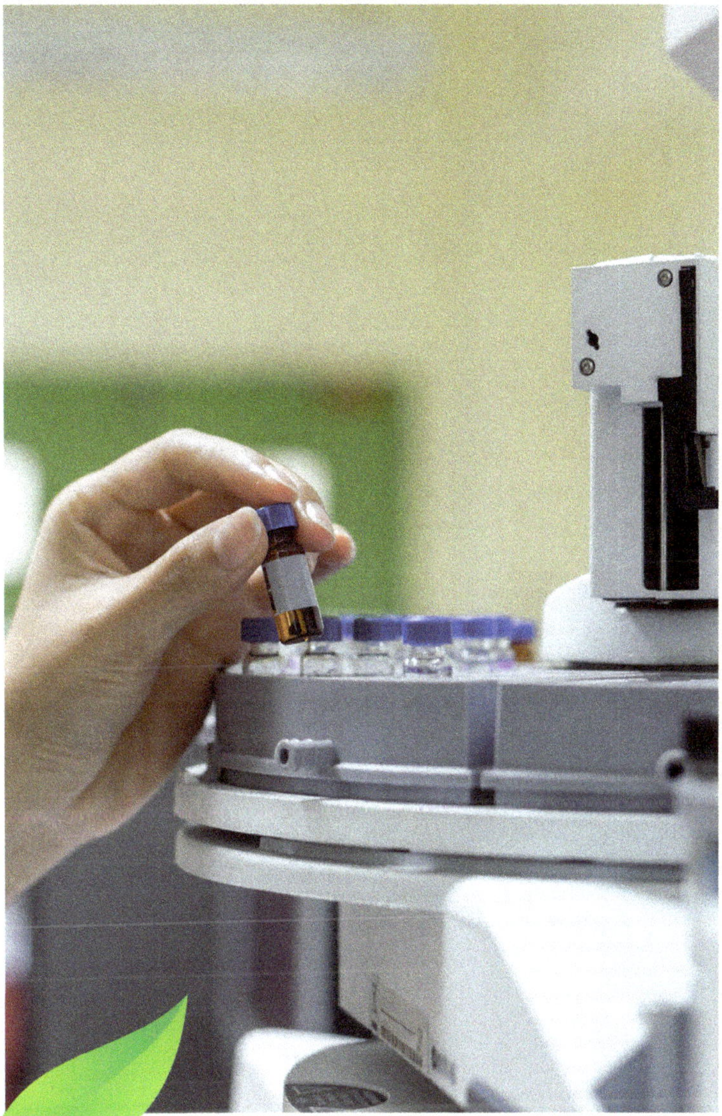

The study of essential oil chemistry relies on various analytical techniques to identify and quantify their components. Gas chromatography (GC) and mass spectrometry (MS) are the most commonly used methods. These techniques allow for precisely identifying individual compounds within an essential oil, providing a detailed chemical profile.

Essential oils are highly concentrated and should always be diluted before use. The right dilution depends on the specific oil, its purpose, and the individual's sensitivity.

CHAPTER SEVEN

Safety and Toxicology in Essential Oils

While offering numerous therapeutic benefits, essential oils must be used with care and knowledge due to their potent nature. Understanding the principles of safety and toxicology is crucial for anyone using essential oils, whether for personal use, aromatherapy, or other applications. This chapter will explore the various aspects of safety, potential toxicological effects, and guidelines for the safe use of essential oils.

Understanding Toxicology

Toxicology is the study of the adverse effects of chemicals on living organisms. In the context of essential oils, it involves understanding how certain components of these oils can cause harm if misused. The toxicity of an essential oil can depend on several factors, including dosage, duration of exposure, method of application, and the individual's age, health, and sensitivity.

Key Toxicological Concepts

- **LD50 (Lethal Dose 50):** This measure indicates the amount of a substance required to kill 50% of a test population. This helps determine the relative toxicity and safe usage levels for essential oils.
- **NOAEL (No Observed Adverse Effect Level):** This is the highest dose at which no adverse effects are observed, providing a benchmark for safe exposure levels.
- **Sensitization and Allergic Reactions:** Repeated exposure to certain essential oils can cause sensitization, leading to allergic reactions. Understanding which oils are more likely to cause such reactions is vital for safe use.

Factors Influencing Safety

Dosage and Dilution

Essential oils are highly concentrated and must be diluted before use. The appropriate dilution can vary depending on the type of oil, its intended use, and the individual's sensitivity. Common carrier oils for dilution include jojoba oil, almond oil, and coconut oil.

Method of Application

- **Topical Application:** When applying essential oils to the skin, dilution is critical to prevent irritation or sensitization. Patch tests are recommended to check for allergic reactions.
- **Inhalation:** While generally safer, some essential oils can cause respiratory irritation, particularly in individuals with asthma or allergies. Proper ventilation and moderate usage are advised.

- **Ingestion:** Internal use of essential oils should be approached with caution and under the guidance of a qualified professional. Not all essential oils are safe for ingestion, and improper use can lead to serious health issues.

Individual Sensitivity

Factors such as age, pre-existing health conditions, and skin sensitivity can influence how a person reacts to essential oils. For example, children, pregnant women, and individuals with certain medical conditions may need to avoid specific oils or use them cautiously.

Common Safety Concerns

Skin Irritation and Sensitization

Essential oils like cinnamon, clove, and oregano are known for their potential to cause skin irritation and sensitization. Always perform a patch test and dilute these oils adequately before topical use.

Phototoxicity

Certain citrus oils, such as bergamot, lemon, and lime, can cause phototoxic reactions when exposed to sunlight or UV light. This can result in skin discoloration, blistering, or burns. It's important to avoid sun exposure after applying these oils to the skin.

Neurotoxicity

Some essential oils, particularly those high in thujone (such as wormwood and sage), can be neurotoxic if used in large amounts or over extended periods. These oils should be used sparingly and with caution.

Pregnancy and Breastfeeding

It's essential to use essential oils with caution during pregnancy and breastfeeding. Some oils can cross the placental barrier or affect breast milk, potentially harming the baby. Oils like peppermint, rosemary, and clary sage should be avoided or used only under professional guidance.

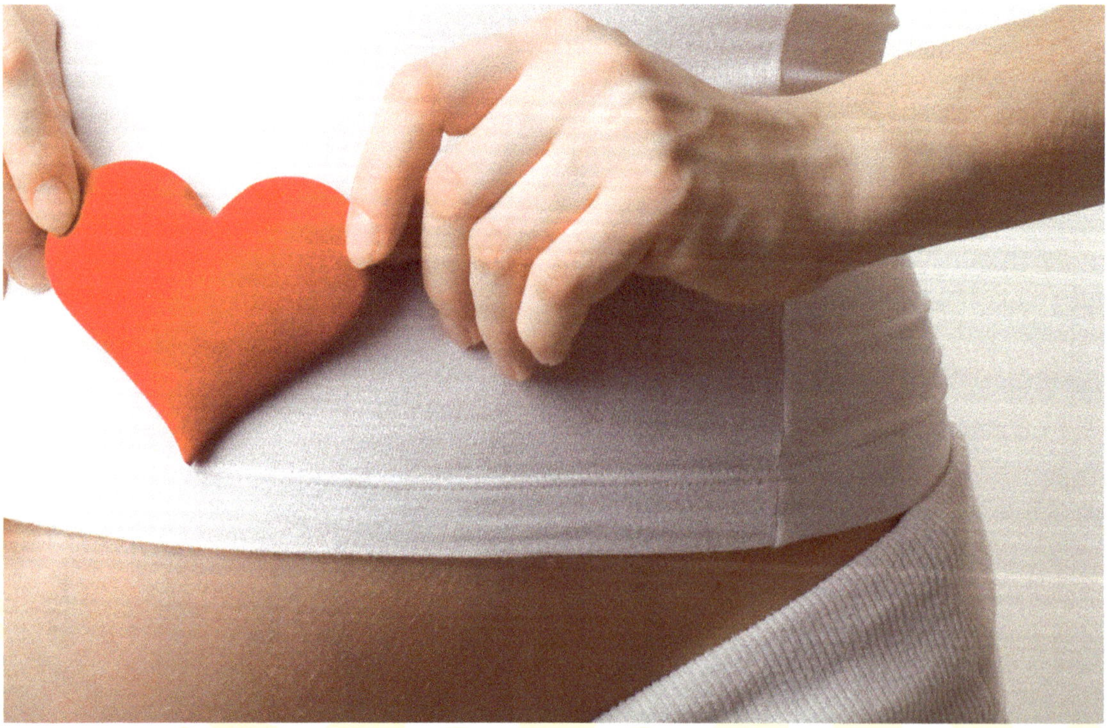

Guidelines for Safe Use

General Safety Tips

- Always dilute essential oils before applying them to the skin.
- Perform a patch test before using a new essential oil.
- Use essential oils in moderation, especially when inhaling or diffusing them.
- Store essential oils in a cool, dark place, away from children and pets.
- Consult a qualified professional before using essential oils internally.

> Essential oils can be toxic to pets, so do your research and use with care.

Special Considerations

- **Children:** Use only child-safe essential oils and dilute them more than for adults.
- **Elderly:** Older adults may have thinner skin and other sensitivities, requiring more caution and lower dosages.
- **Pets:** Some essential oils can be toxic to pets, especially cats and dogs. Research pet-safe oils and avoid using diffusers in confined spaces where pets are present.

The potent nature of essential oils necessitates a thorough understanding of their safety and toxicology. By adhering to proper usage guidelines, recognizing potential hazards, and respecting individual sensitivities, essential oils can be used effectively and safely to enhance well-being. This chapter aims to provide a comprehensive guide to ensure that the therapeutic benefits of essential oils are harnessed without compromising safety.

Research on LD50 of Essential Oils

Understanding the toxicological profile of essential oils is crucial for their safe and effective use. One key measure used in toxicology to assess the relative toxicity of substances, including essential oils, is the LD50, or Lethal Dose 50. This value represents the dose required to kill 50% of a test population, typically laboratory animals like rats or mice. It provides insight into the potential risks associated with various essential oils. Research on the LD50 values of essential oils helps determine safe usage levels, guiding practitioners and users in making informed decisions about dosage and application. By exploring the LD50 values of common and less common essential oils, we can better understand their toxicity and establish guidelines for their safe use in aromatherapy and other applications.

Common Essential Oils:

- **Eucalyptus Oil:** When administered orally, research has indicated an LD50 value for eucalyptus oil in rats ranging from 2.48 to 3.51 g/kg. This suggests moderate toxicity.
- **Peppermint Oil:** Studies have shown the oral LD50 for peppermint oil in rats to be approximately 2.426 g/kg, indicating moderate toxicity.
- **Tea Tree Oil:** The oral LD50 for tea tree oil in rats is around 1.9 g/kg, reflecting relatively higher toxicity compared to some other essential oils.

Knowing the LD50 of essential oils helps set safe dosage guidelines.

Less Common Essential Oils:

- **Wintergreen Oil:** Known for its high methyl salicylate content, it has an oral LD50 value in rats of about 1.2 g/kg, indicating significant toxicity.
- **Clove Oil:** With an oral LD50 value in rats of approximately 1.375 g/kg, clove oil is considered to have moderate to high toxicity.

Implications of LD50 in Essential Oil Use

1. **Guidelines for Safe Use:** Understanding the LD50 of essential oils allows for developing guidelines on safe dosage levels, particularly when used internally or topically.
2. **Risk Assessment:** LD50 values help assess the risk of acute toxicity and guide the recommendations for safe handling and usage.
3. **Formulation of Products:** When formulating products that include essential oils, knowing their LD50 values ensures that concentrations remain within safe limits to avoid adverse effects.

Studies and Resources

1. **Tisserand and Young's "Essential Oil Safety":** This comprehensive book provides detailed information on numerous essential oils' toxicity, including LD50 values.
2. **PubMed and Research Articles:** Various studies published in scientific journals accessible via databases like PubMed explore the toxicological profiles of essential oils, including their LD50 values.
3. **Regulatory Agencies:** Organizations such as the Environmental Protection Agency (EPA) and the European Medicines Agency (EMA) often include toxicological data, including LD50, in their

safety assessments of essential oils and related substances.

Practical Application

For practitioners and users of essential oils, understanding the LD50 values is crucial for:

- Determining safe dilution ratios.
- Identifying oils that should be used cautiously or avoided in certain populations (e.g., children, pregnant women, pets).
- Making informed decisions about using essential oils in various therapeutic and personal care applications.

> *Identify oils to avoid for children, pregnant women, and pets.*

In summary, LD50 values are a valuable tool in assessing the toxicity of essential oils and ensuring their safe and effective use. By incorporating this knowledge into practice, we can better harness the therapeutic benefits of essential oils while minimizing potential risks.

Exploring the science behind essential oils at Aroma Hut Institute, where we uncover the chemistry behind their therapeutic power.

Understanding potential antagonism between essential oils before blending requires both theoretical knowledge and practical experience.

CHAPTER EIGHT

Advances and Innovations in Essential Oil Chemistry

The field of essential oil chemistry is continuously evolving, with new research uncovering diverse therapeutic properties and applications. Recent studies have explored lesser-known essential oils' antimicrobial, anti-inflammatory, and antioxidant activities. For example, research on essential oils from exotic plants like *Litsea cubeba* and *Zanthoxylum armatum* has revealed promising bioactivities, prompting further investigations into their potential uses in medicine and cosmetics.

Additionally, advancements in analytical techniques such as gas chromatography-mass spectrometry (GC-MS) and nuclear magnetic resonance (NMR) spectroscopy have enabled more precise identification and quantification of essential oil components. This has led to the discovery of novel compounds and the elucidation of their mechanisms of action, providing a deeper understanding of how essential oils exert their effects.

> Technological advancements have greatly enhanced the extraction and analysis of essential oils.

Technological Innovations in Extraction and Analysis

Technological advancements have significantly improved the extraction and analysis of essential oils. Traditional methods like steam distillation and cold pressing are now complemented by innovative techniques such as supercritical fluid extraction (SFE) and microwave-assisted extraction (MAE). These methods offer higher efficiency, reduced extraction times, and the ability to preserve sensitive compounds that might degrade under conventional conditions.

Supercritical fluid extraction, in particular, uses supercritical carbon dioxide as a solvent, which is both environmentally friendly and highly effective at extracting essential oils without leaving harmful residues. Microwave-assisted extraction utilizes microwave energy to heat plant materials rapidly, leading to faster extraction and higher yields.

In the realm of analysis, integrating advanced software and machine learning algorithms has enhanced the interpretation of complex data from GC-MS and NMR. These technologies enable the identification of subtle variations in essential oil composition, facilitating quality control and authenticity verification.

Sustainable Practices and Green Chemistry

Sustainability is a growing concern in the production of essential oils. The demand for essential oils has led to overharvesting and environmental degradation in some regions. Researchers and producers are adopting sustainable practices and principles of green chemistry to address these issues.

Cultivating aromatic plants using organic farming methods, crop rotation, and integrated pest management helps maintain soil health and biodiversity. Developing biotechnological approaches, such as plant tissue culture and genetic engineering, also offers the potential for sustainable production of essential oils without overexploiting natural resources.

Green chemistry principles are being applied to minimize waste and reduce the environmental impact of essential oil extraction and processing. This includes the use of renewable solvents, energy-efficient extraction methods, and the recycling of by-products. These practices contribute to environmental conservation and enhance the quality and safety of essential oils.

Future Directions in Essential Oil Chemistry

The future of essential oil chemistry holds exciting possibilities as researchers continue to explore new applications and refine existing ones. One promising area is the use of essential oils in nanotechnology. Encapsulating essential oils in nanoparticles can enhance their stability, bioavailability, and controlled release, opening up new avenues for drug delivery and topical applications.

Another emerging field is the integration of essential oils with modern medicine. Collaborative research between chemists, pharmacologists, and clinicians is investigating the synergistic effects of essential oils with conventional drugs, aiming to develop more effective and personalized therapeutic approaches.

Furthermore, the increasing interest in personalized wellness drives the development of customized essential oil blends based on individual genetic profiles and health conditions. Advances in genomics and metabolomics will enable more precise tailoring of essential oil therapies to meet specific needs.

In conclusion, the dynamic field of essential oil chemistry continues to evolve with emerging research, technological innovations, sustainable practices, and future directions that promise to enhance essential oils' efficacy, safety, and environmental impact. As scientists and practitioners collaborate, the potential for essential oils to contribute to health, wellness, and ecological sustainability will only continue to grow.

The future of essential oil chemistry lies in deeper exploration of their molecular interactions and therapeutic potential.

"Using essential oils sustainably enhances their benefits while minimizing environmental impact.

CHAPTER NINE

Conclusion

Essential oils testify to nature's intricate design, boasting various chemical compounds contributing to their diverse therapeutic properties. From the fragrant lavender to the invigorating peppermint, each essential oil is a unique blend of terpenes, aldehydes, ketones, esters, and more, working synergistically to produce their characteristic effects.

Throughout this book, we have delved into essential oils' chemical families and components, exploring their interactions and benefits. The rich complexity of essential oils is not merely a subject of academic interest; it forms the basis of their practical applications in aromatherapy, medicine, and personal care. Understanding the chemistry behind these oils enhances our appreciation of their power and potential, guiding us in their effective and safe use.

The field of essential oil chemistry is dynamic and evolving, with new research and technology revealing insights and possibilities in both lesser-known oils and advanced extraction methods.

The Ongoing Journey of Chemical Exploration

The field of essential oil chemistry is dynamic and ever-evolving. Advances in research and technology continuously unveil new insights and possibilities. Emerging research has expanded our knowledge of the biological activities of lesser-known essential oils, while technological innovations have revolutionized extraction and analysis methods.

Our journey through this book has highlighted the importance of sustainable practices and green chemistry in the production and application of essential oils. As we strive to meet the growing demand for these natural products, it is crucial to balance this with environmental stewardship and ethical considerations.

The ongoing journey of chemical exploration promises exciting developments in the future. The potential applications are vast and varied, from integrating essential oils with nanotechnology to personalized wellness solutions. This journey is far from over, and each discovery brings us closer to unlocking the full potential of essential oils.

Encouraging Further Study and Application

As we conclude this book, it is important to recognize that our understanding of essential oils is still in its infancy. There remains a vast expanse of knowledge to be discovered and many more applications to be explored. The complexity of essential oils offers endless opportunities for research, innovation, and practical application.

For those in aromatherapy, chemistry, or anyone with a passion for natural health, I encourage you to continue studying and experimenting with essential oils. Whether you are a researcher seeking to uncover new therapeutic properties, a practitioner aiming to enhance client care, or an enthusiast exploring the world of essential oils for personal wellness, your contributions are invaluable.

The journey of essential oil chemistry is a collaborative effort built on individuals' curiosity, dedication, and creativity worldwide. By sharing knowledge, conducting rigorous research, and applying best practices, we can continue to advance this fascinating field.

In conclusion, essential oils represent a remarkable intersection of nature, science, and wellness. Their rich complexity and myriad benefits inspire us to explore, innovate, and apply their potential in ways that enhance our health and well-being. As we move forward, let us embrace the ongoing journey of chemical exploration and encourage further study and application, ensuring that essential oils remain a vibrant and integral part of our lives.

Essential oils merge nature and science, inspiring ongoing exploration to enrich our health and well-being.

”

Understanding potential antagonism between essential oils before blending requires both theoretical knowledge and practical experience.

www.ingramcontent.com/pod-product-compliance
Ingram Content Group UK Ltd.
Pitfield, Milton Keynes, MK11 3LW, UK
UKHW060223061026
15625UKWH00007B/463

9 798987 246467